EAST BAY HILLS

• A BRIEF HISTORY •

AMELIA SUE MARSHALL

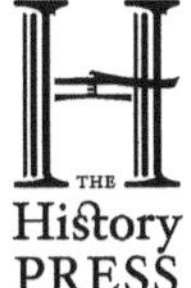

Published by The History Press
Charleston, SC
www.historypress.net

Cover images: *Top front*: Riders from the San Ramon Valley Horsemen on the East Ridge Trail in Redwood Regional Park, circa 1975. Info provided by Per Nielsen, George Wagnon, Bryan Hodges, Mildred Dalgleish and Esperance (Es) Anderson. *Courtesy of East Bay Regional Parks archives*; *bottom front*: The Palo Seco Mill around 1910. *Courtesy of the Oakland Public Library Oakland History archives*; *top back*: *Photo by the author*; *back inset*: Dr. Walter Littlejohn and companions. *Courtesy of East Bay Regional Parks archives.*

First published 2017

Manufactured in the United States

ISBN 9781467137256

Library of Congress Control Number: 2017945010

Notice: The information in this book is true and complete to the best of our knowledge. It is offered without guarantee on the part of the author or The History Press. The author and The History Press disclaim all liability in connection with the use of this book.

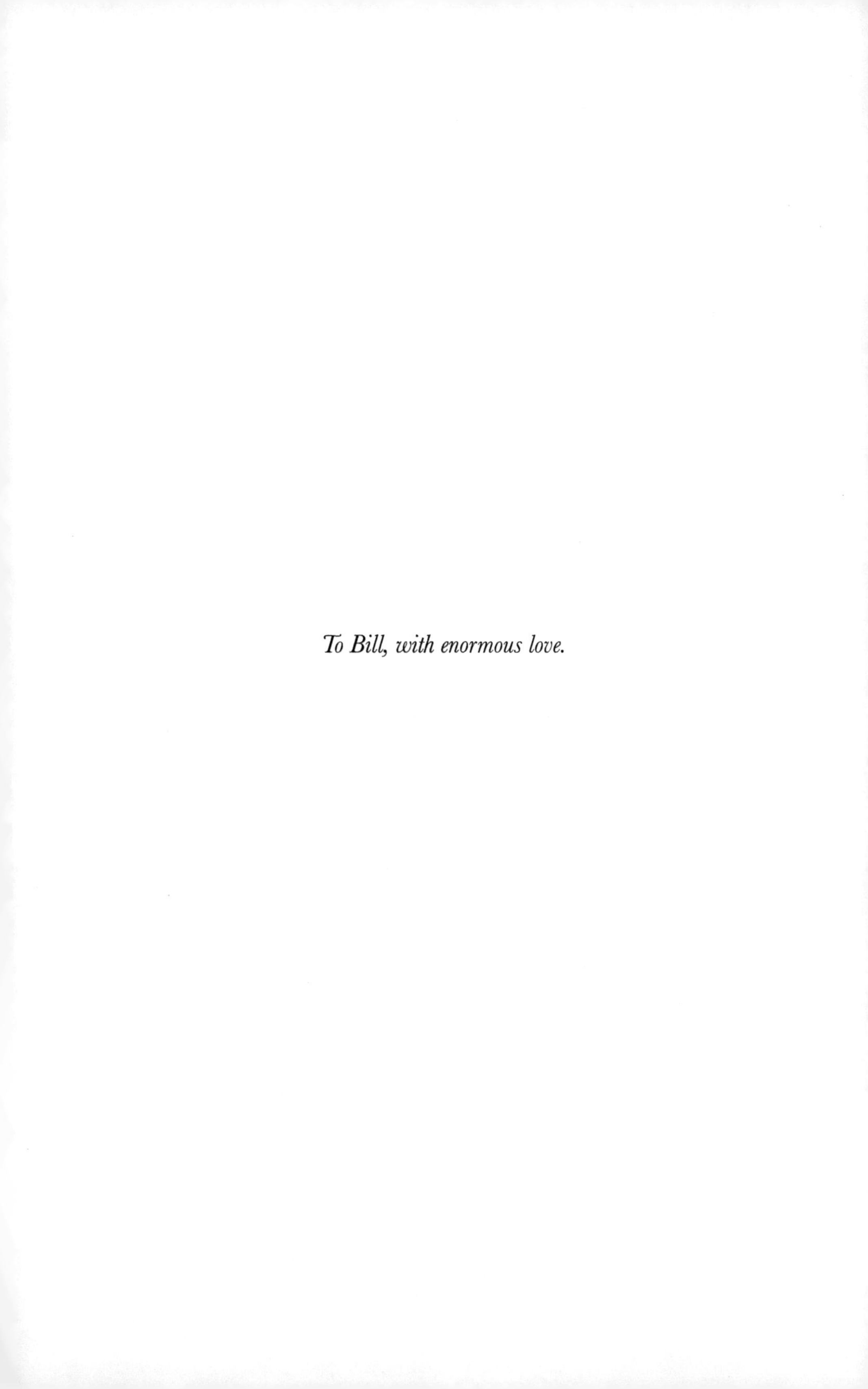

To Bill, with enormous love.

CONTENTS

ACKNOWLEDGEMENTS

We Oaklanders try to celebrate our cultural heritage in a style that is audacious, righteous and open-hearted. So, let's start with a salute to Dennis Evanosky, dean of the local history writers, for his unstinting generosity in sharing the knowledge he has accumulated over many years. Following along in the open-source spirit are writers Annalee Allen, Gene Anderson, Jacqueline Beggs, Ed Clausen, Eleanor Dunn, Richard Langs, Erika Mailman, Doris Marciel, Malcolm Margolin and Richard Schwartz.

Our secret society at the East Bay Regional Park District archives, under the "adult supervision" of the redoubtable Brenda Montano, has been the wellspring of ideas for this book. The lovely and talented Jerry Kent, Ned MacKay, Paul Miller, Anne Rockwell, Beth Stone, Leslie Stoup, Ron Ucovich and Bill Vidor have been my source of nurturance and inspiration for several years.

The late Lloyd Graham Donaldson, known to everyone as Lloyd Graham, graciously spent many hours sharing his reminiscences about his long lifetime in the East Bay redwood forest. We lament his passing before he could see this book in print.

Regional Parks Director Dee Rosario and the wonderful Robyn Krieger are a constant source of information, energy and fun. Past and present East Bay Regional Park District (EBRPD) staff who have contributed to this work include Michael Charnofsky, Janet Gomes, Annie Kenny, Justin Neville, John Nicoles, Joe Sullivan and Dave Zuckermann.

Others who shared vital information include Judi Bank, Larry Brickell, Rex Burress, Bob Cooper, Jim Covel, Heather Galanis, Kendall Langan, Dorothy Londagin, Martin Mataresse, Kimra McAfee, Dwayne McCosker, Mary McCosker, Erik Olafsson, Cecelia Peña, Greg Pratt, Dave Radlauer, Bob Schultz, Esperanza Pratt Surls, Stuart Swiedler, John Whatley and Rich Wirkkala.

This work has been enriched with information generously shared by staff and volunteers from local historical organizations: Holly Alonso, of the Antonio Peralta Hacienda Museum; John Christian, of the Hayward Area Historical Society; Elsie Mastick at the Moraga History Center; and Chris Doan, archivist for the Roman Catholic Archdiocese of San Francisco.

Oakland librarians Dorothy Lazard, Steve Lavoie and Kathleen DiGiovanni are invaluable resources to our community, as are planners Betty Marvin and Gail Lombardi at the Oakland Cultural Heritage Survey. Peter Hanff and Michael Maire Lange at the Bancroft Library and Allan Fisher at the Western Railway Museum have been very helpful in locating historic photographs.

Cartographer Alan Imler worked tirelessly to perfect our maps.

Editor Deborah Dunster has spared me much embarrassment through her exacting attention to the prose.

Graphic designer Naomi Schiff performed magic tricks with obscure illustrations to reveal hidden mysteries.

Thanks to Stan Dodson for introducing me to The History Press team. Thanks for your unbounded service in local parks—from chain-sawing downed trees to leading nature hikes.

Any shortcomings of this work are my sole responsibility.

This book might never have been written if it weren't for Neil Stollman MD, the world's best doctor and a truly delightful human being.

My husband and best friend, Bill Imler, PhD, has provided ongoing encouragement, nightly five-star dinners and daily acts of kindness throughout my process of researching and writing this book. We are grateful to our beloved offspring, Alan and Rosalie, who bring us joy each day.

INTRODUCTION

These are interesting facts that we may be pardoned for producing.
—History of Alameda County, *by M.M. Wood, 1883*

The East Bay hills hold the only inland forest of coastal redwood trees within a major metropolis. Its old-growth progenitor trees may have been the tallest anywhere. Sea captains southbound from the Bering Strait gazed upon them to set a course through San Francisco Bay.

Saclan Bay Miwok people of the inland valleys and Ohlone people of the coastal plains have lived here for centuries, and their cultures endure.

Yankee bullwhackers and Portuguese farmers built homesteads in the fragrant redwoods. They founded schools and held elections. They organized fire protection districts.

Rowdy cowboys whooped it up in the saloons of Redwood Canyon. Rustlers drove away other people's livestock. Mounted rangers from the redwoods brought them to justice. The Wild West lived large in the hidden canyons.

Civil servants and university professors imagined a swath of parkland stretching twenty miles along the hills. Lawyers and union leaders conferred for years until the dream became reality. The labor armies of the New Deal rescued families from poverty, with Italian stonemasons building shelters and barbecue ovens for the public to enjoy.

East Bay voters chose to tax themselves for parkland twice—in 1934, during the Great Depression, and again in 2008, with the Great Recession underway.

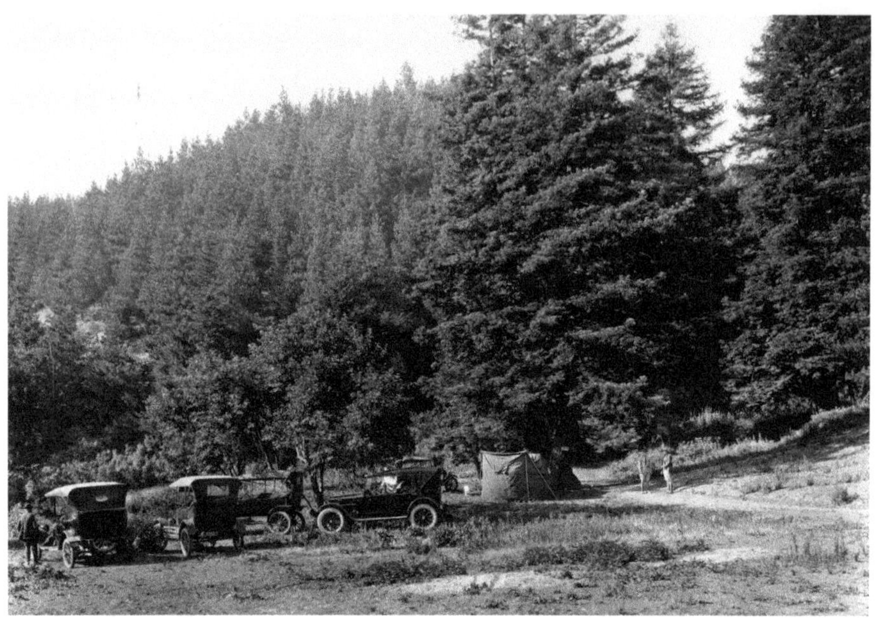

Crews from the Works Progress Administration (WPA) arrive in Redwood Regional Park, circa 1936. *Courtesy East Bay Regional Park District archives.*

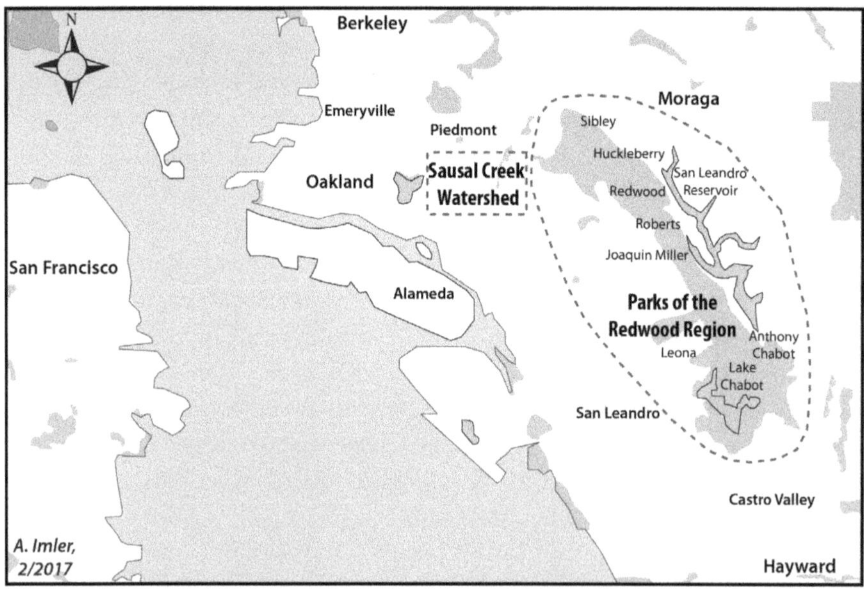

The East Bay Hills redwood region—the only inland forest of *Sequoia sempervirens*—is found in the hills between Oakland and Moraga. *Courtesy Alan Imler Cartography.*

Who planted the olive trees on the west slope of Redwood Peak? Who built the stone stairs that once led to the Redwood Hunting Lodge? For how many years did Ruby's bordello serve her rustic clientele?

There's more to the stories than you may have expected!

Chapter 1

NATIVES

As the eighteenth century was drawing to a close, opposition factions in tribal villages continued to resist Spanish intrusions into the East Bay hills…among them were escaped Christian and non-Christian Saclans, allied with some Jalquins. But…small groups of forty or fifty warriors, armed only with bows, lacked the power to protect their borders against Spanish invaders.
—Randall Milliken, A Time of Little Choice, *2009*

Indigenous Peoples and Their East Bay Homelands

The East Bay redwood forest is surrounded by the traditional homelands of three distinct tribal groups: the Saclan, the Huchiun and the Jalquin-Yrgin.[1] While all were multilingual, the primary ancestral language of the Saclan was Bay Miwok. The Huchiun and Jalquin-Yrgin spoke East Bay Ohlone/ Costanoan[2] as their primary language.

The forest itself is in an area where tribal territories likely overlapped. It is the Jalquin-Yrgin or Saclan people who are believed to have been the primary residents near the redwood trees. In general, villages were located on major creeks that had reliable year-round water.

Within the East Bay redwood forest, subsequent logging activity has obliterated practically all evidence of Native habitation. Mortar rocks exist in a few remote locations. Incidentally, petrified redwood was present at Moraga sites, indicating that the redwood forest extended at least as far as Las Trampas in centuries long ago.[3]

Native people lived in harmony with nature for many centuries before the arrival of Europeans. Much knowledge of their traditional way of life has survived, passed on from generation to generation. But much has also been lost. The missionaries who accompanied the Spanish explorers aimed to extinguish the tribal cultures, seeing this as essential in achieving their goal of Christianizing the Natives.

Beyond the traditions that have been kept alive within the contemporary Ohlone community, much of what is known about the ancestral East Bay indigenous people comes from reports by early Spanish explorers and mission records.

"*Gente de los Palos Colorados*," people of the redwoods, as they were identified in mission records, may have come from East Bay locations far from the hills. As various priests kept records at Mission San José, different conventions were followed in recording tribal data from 1797 to 1833.

Traditional Ways of Life

Before 1769, Native people had a varied and nutritious diet, using a wide variety of local plants. Protein sources were also plentiful, as there was abundant fish and game.

East Bay women gathered acorns and seeds from flowering plants to make into soups (*atole*) and savory cakes (*pinole*), steaming some in baskets. Acorns were carefully prepared by grinding the savory kernels into a flour that was used in various dishes. Acorn soup may have been cooked in baskets, using hot stones.

Men hunted deer, elk and bear using bows and arrows. Arrows were made from straight tule rods, with arrowheads of quartz or obsidian. Trout and salmon were taken from creeks using loosely woven baskets. Small game was caught using snares.

Native people managed the understory of the redwood forest and oak woodlands with small fires. This cleared brush from under the ancient trees. People lived in dome-shaped dwellings of branches thatched with tule. Granaries and nets were designed to protect foodstuffs from pests.

Traditional Native people had rich spiritual and ceremonial lives, with legends passed orally between generations. It is believed that ceremonial regalia, as well as clothing styles, differed greatly from one tribal group to another.

Of the people who lived in the East Bay redwoods, little is known of their specific practices. Could they have used redwood saplings in the construction of their dwellings and baskets in the Pacific coast redwoods, as was done by the people of Ano Nuevo and Santa Cruz?

For a complete discussion of everyday life among the ancestral Ohlone/Costanoan people, please refer to the *Ohlone Curriculum* by Beverly Ortiz, PhD, available on the East Bay Regional Parks website (www.ebparks.org).

The East Bay Ohlone/Costanoans: Huchiuns

The Huchiun people lived on the flatlands and foothills of today's Oakland and Berkeley. They built a sweat lodge on Temescal Creek, between today's Montclair and upper Rockridge districts of Oakland. The lodge site was flooded when Lake Temescal was built by Anthony Chabot in 1869.

The mouth of Temescal Creek lies opposite the Golden Gate to the Pacific. Large marine creatures such as whales, sharks and seals came to shore. Indigenous people traveled from afar to enjoy this abundant seafood. Shellfish were probably gathered from boats crafted from tule reeds.

The Huchiun people lived in widely spaced villages. In 1776, explorer Juan Bautista de Anza, missionary Francisco Palou and sea captain Jose Canizares reported seeing villages with between forty and one hundred inhabitants. Canizares also observed one large village near the Carquinez Strait with four hundred people present. Perhaps this was an intertribal gathering.

The Jalquin-Yrgin People

Anthropologist Randall Milliken carefully studied records from Mission San Francisco ("Mission Dolores"), Mission San José (in today's Fremont) and Mission Santa Clara to determine locations of tribal homelands. While these records refer separately to the Jalquin and Yrgin tribes, Milliken determined that these peoples were highly intermarried. Milliken mapped the traditional homelands of the Jalquin people to today's San Leandro, along San Leandro Creek, and the southern part of today's Oakland. The Yrgin lived in today's Castro Valley and in the vicinity of Anthony Chabot Regional Park.

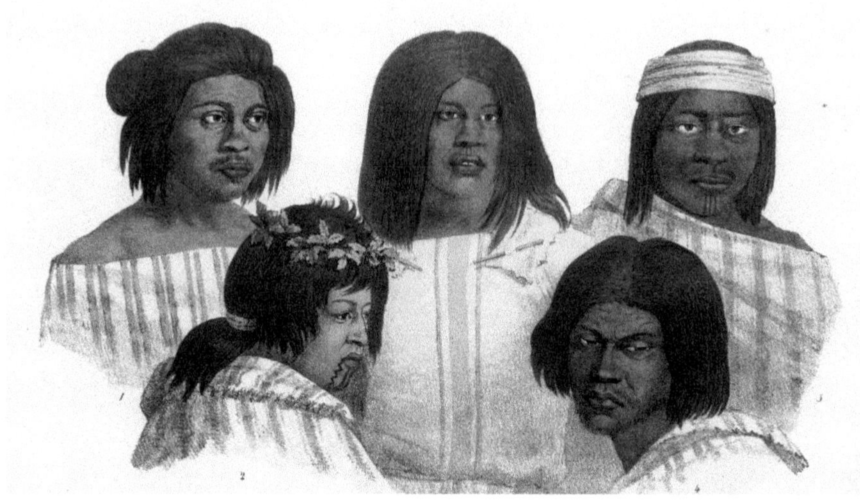

Individuals from different tribal groups present varied features. Ships' artist Ludwig Choris sketched these Natives at Mission Dolores in San Francisco in 1816. *Courtesy Bancroft Library.*

The Saclan People

Before to the mission era, the Saclan people occupied inland valleys, from San Pablo Creek in the north through Walnut Creek, Lafayette, Orinda and Moraga; east through the San Ramon Valley; and south through Sunol. Mission records stated that the Saclan villages of Jussent and Guquigmu were located within a few miles of each other, probably in the area of Lafayette or Moraga.

The last documented Saclan resident of Moraga continued to live in Indian Valley after 1797, after her fellow tribal members had gone to Mission San José.[4]

LANGUAGES SPOKEN BY EAST BAY NATIVES

Europeans observed early on that Natives could understand the languages spoken by neighboring tribes. Sailors on the ship *San Carlos* observed that Costanoan-language speakers in the East Bay could understand the words the sailors had learned from Rumsen Costanoan speakers in Monterey.

The first European scholars to take a serious interest in East Bay indigenous languages were two Franciscan friars. Francisco Palou became the founder of Mission San Francisco (Mission Dolores) in 1776. Felipe Arroyo de Cuesta, who arrived in California in 1808, traveled among several missions, compiling notebooks on Saclan and Huchiun (Chochenyo), the languages of the East Bay redwood people.

Beginning in 1901, Alfred L. Kroeber and his colleagues at the University of California–Berkeley sought to classify Northern California indigenous people according to the primary language they spoke.

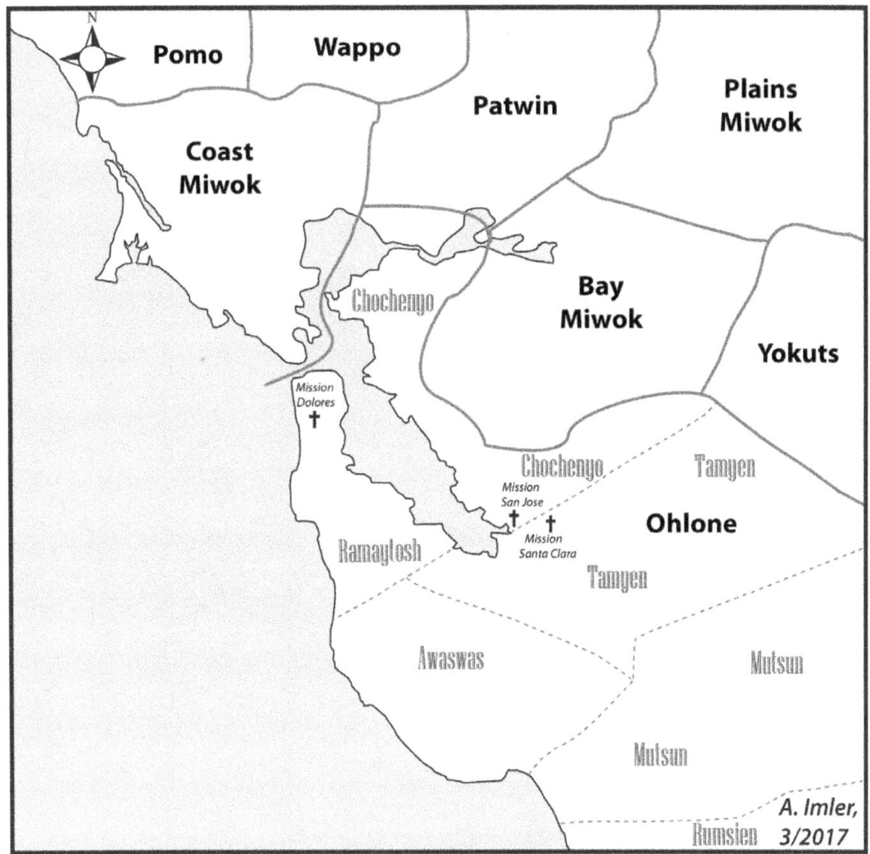

Indigenous People of the Bay Area:
Primary Languages Spoken, Regional Dialects

California indigenous people have been grouped by anthropologists according to their principal language. *Courtesy Alan Imler Cartography.*

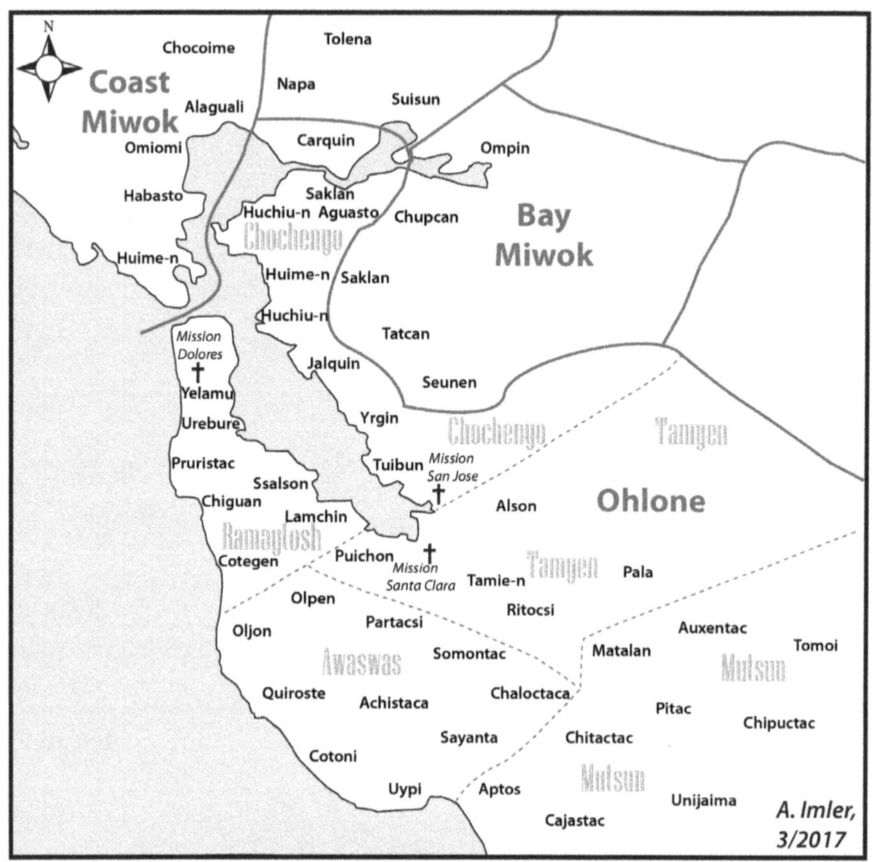

Indigenous People of the Bay Area:
Some of the Tribal Names in Ohlone and Miwok Regions

Before the mission era, Natives lived in many small tribal groups. Tribe names generally did not correspond to languages and dialects spoken. Based on research by Randall Milliken. *Courtesy Alan Imler Cartography.*

John Peabody Harrington, working for the federal Bureau of American Ethnology, collected information and recorded Native speakers on wax cylinders between 1921 and 1929. Maria de los Angeles (Angela) Colos, of Pleasanton, began teaching Harrington and others the East Bay Costanoan language in 1921. José Guzman continued the instruction in 1930. Harrington named this language "Chochenyo."

In 1955, Madison Beeler of UC-Berkeley caught some of Kroeber's errors in identifying the homeland and language relationships of the Saclan. Beeler identified the Saclan language as being closely related to the Miwok

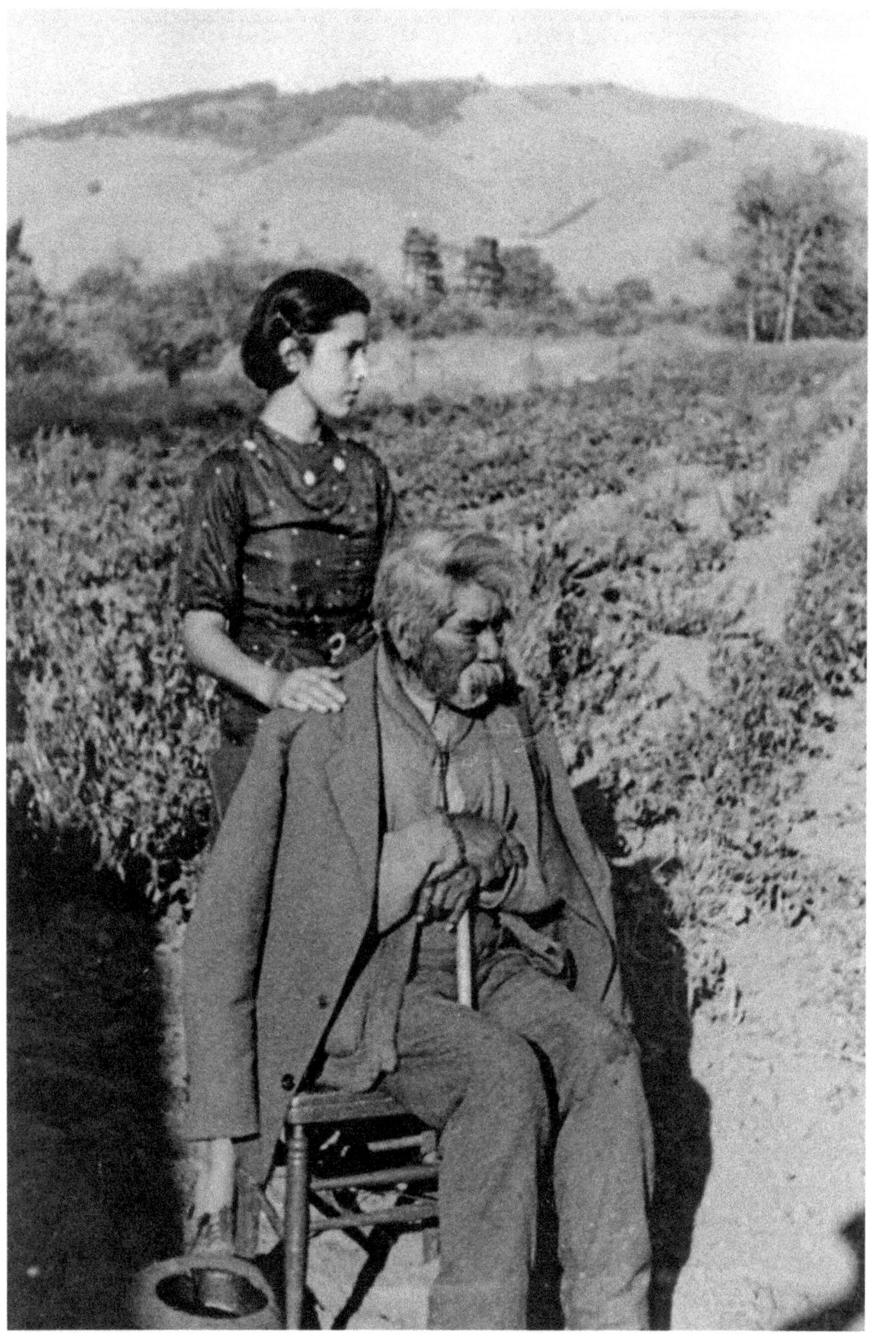

José Guzman, shown here with an unidentified girl, taught the Chochenyo dialect of the Ohlone language to Smithsonian anthropologist John Peabody Harrington in the 1930s. *Courtesy Antonio Peralta Hacienda Museum archives.*

languages of the Sierra people. Thus, he gave the name Bay Miwok to the Saclan tongue.

Beeler characterized the Costanoan language as having three branches: the northern group, including Chochenyo; a southern group, including Mutsun (San Juan Bautista) and Rumsen (Monterey); and Karkin (Martinez), which is a unique branch of Costanoan.

In 1976, Richard L. Levy of UC-Berkeley and his colleagues coined names for the indigenous languages, trying to use more culturally appropriate Native words rather than the mission names applied by Kroeber. The name Ohlone, derived from the name of the Oljon tribe of Pescadero and San Gregorio, is now in widespread use for the Costanoan people.

In 1995, Randall Milliken began publishing his studies of mission records. Interestingly, he found that the name endings of women from the Jalquin tribe followed the conventions of the Bay Miwok language, but those of Yrgin females followed the Chochenyo Ohlone language.

The Chochenyo Language

Linguists today identify six languages of the Ohlone/Costanoan family: San Francisco Bay Costanoan, Karkin, Awaswas, Mutsun, Rumsen and Chalon. Within the San Francisco Bay Costanoan, there is an East Bay dialect: Chochenyo, the language of the Huchiun people. And within Chochenyo, there is even an Oakland dialect.

Thanks to the work of Angela Colos, José Guzman and John Peabody Harrington, much of the Chochenyo language has been preserved. A revival of Chochenyo is now underway among Bay Area Natives.

The Bay Miwok Language

Except for about one hundred words and phrases, the Bay Miwok language has been lost. In 1821, Crispio Jahuoesia taught these words to the Franciscan missionary Felipe Arroyo at the Mission Dolores *rancheria*. The priest made note of these words and phrases, spelling them as they sounded to a Spanish-speaker. In an 1821 commentary, Arroyo described Saclan as a "graceful" language.

Natives Meet Europeans

Earliest Encounters

In initial encounters between the Spanish cavalry and the Native people, the newcomers were greeted warmly and, in some cases, with awe and wonder. Expedition leaders Pedro Fages and Juan Crespí and mariner Jose Canizares recounted how their hungry troops received gifts of food—*pinoles*, acorn and seed cake that tasted like roasted hazelnut and cooked fish. The Europeans gave the Natives glass beads and other items in exchange.

People of the village on San Pablo Bay told Canizares in 1775 that they had seen men on horseback before. This must have been the Fages-Crespí expedition of 1772. "Without doubt, horses were initially the most intimidating symbol of the Spaniards' extraordinary, possibly supernatural power," observed Randall Milliken.

Initially, people of the redwoods who entered the mission system went to Mission San Francisco, using their tule reed boats to cross the bay. Missionary Vicente Santa Maria, diarist on Canizares's ship, the *San Carlos*, recorded the names of Huchiun-Aguastos men who came to visit the ship: Capitan Sumu, second chieftain Jausos, Supitacse, Mutuc, Logeacse and Xacacse. Records show that several of these men and their family members were baptized at Mission San Francisco (Mission Dolores).

The first large groups of Huchiun went to Mission San Francisco in the fall of 1794. The first baptisms of Saclan people were recorded that same year. After Mission San José was built in 1797, East Bay Natives went there to live instead.

Natives Work on the Ranchos

After the founding of Mission Dolores in 1776 and Mission San José in 1797, many Native people had moved to the mission *rancherias*, leaving vast swaths of unoccupied land. Statewide, the missions capitalized on this wealth of real estate through cattle ranching. This not only provided a way for the residents to sustain themselves and the church but also generated export commodities in the form of tallow and hides, in an economy that was largely based on the barter system. Value was added through the unpaid labor of Native men, who worked as cowboys, and women, who performed every domestic task imaginable.

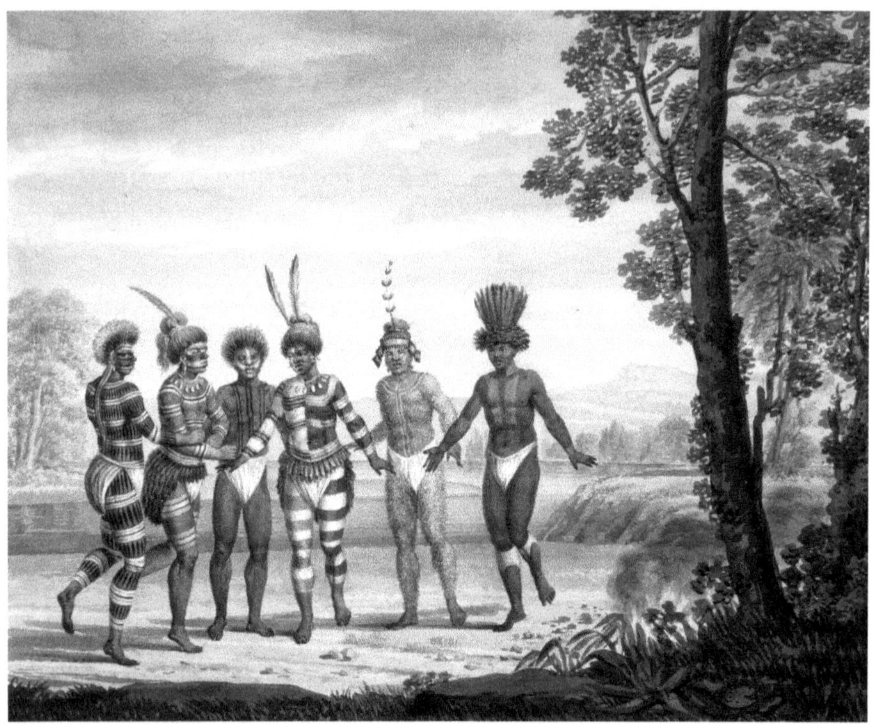

Ohlone men were given time off from their hard labor at Mission San José to demonstrate a ceremonial dance for visitors. Wilhelm G. Tilesius von Tilenau is believed to be the artist who depicted the dancers at Mission San José sometime between 1803 and 1807. *Courtesy Bancroft Library.*

One local example of Native cowboys working mission cattle in the early nineteenth century was *El Rancho San Ysidro de los Juchiunes.* This was a cattle ranch in the Richmond–San Pablo region of the East Bay to supply Mission Dolores. The *vaqueros* were Huchiun men.

Organized Resistance to the Mission System

From the perspective of the Spanish missionaries, it was their duty to bring the Natives into the protective embrace of the church. They were taught to abandon traditional ways. They were to provide a labor force for the missions. The details of the bargain were not initially spelled out. But once a person had been baptized, the Spanish expected him or her to accept

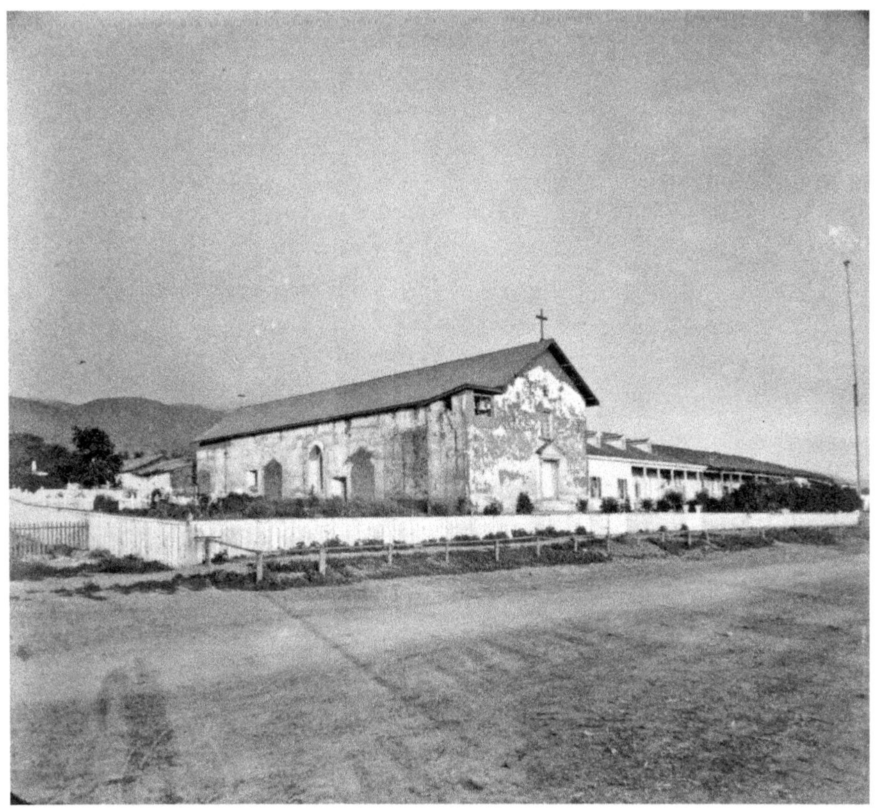

Natives from many tribes, including the people of the redwoods, *Gente de los Palos Colorados*, lived and died at Mission San José. *Courtesy Society of California Pioneers.*

them as their masters. Once *neofitos* entered the mission system, they were forbidden to leave. The *padres* did not hesitate to send troops to bring back runaways (*cimarrones*).

The Chimenes Incident

After an epidemic, likely typhus, caused many deaths at Mission San Francisco in March 1795, a group of Saclan converts secured permission to return to their homelands for a visit (*paseo*). What followed was recorded in detail in the diary of Monterey governor Diego Borica. When the Saclan converts had failed to return by April 27, missionary Antonio Danti sent a group of fourteen Christianized men to pursue them to a ceremonial dance at a village the Spanish called Chimenes Rancheria. (Milliken suggested that this was probably the village of Tcimenukme near today's Napa.) When eight Mission San Francisco men confronted the runaways outside the dance house, a battle ensured. Seven of the pursuers died. Only Oton, a Huchiun pursuer from the mission, survived to tell the tale.

Rounding Up Runaways

Two years later, having resolved to build the new Mission San José, missionaries sent one Costanoan and one Bay Miwok interpreter, with an escort of three soldiers, to Native villages on the Fremont plains to recruit a construction crew. There were few, if any, volunteers. The church men observed that at least two hundred of the village inhabitants were runaways from Mission San Francisco.

Around the same time, on July 20, 1797, authorities at Mission San Francisco sent a party of thirty *neofito* troops northward to round up runaways. Their *capitán* was Raymundo El California, a Native from Baja. The men crossed the bay in tule balsa boats, apparently landing at San Pablo Bay.

Raymundo's troops discovered three Huchiun villages with some of the escapees—men, women and children—present. The Huchiun people fought the troops to avoid being recaptured. The troops retreated to their boats, pursued by the Huchiuns.

Spanish Cavalry Attack on the Saclan

Meanwhile, back at Mission San José, Sergeant Pedro Amador had arrived by July 7, 1797. In what may have been a tragic cross-cultural misunderstanding, Amador had heard that the Saclan men were manufacturing arrows with intent to go to war. This may have merely been a seasonal gathering of men making the year's arrows for hunting, as the tule reeds were ripe for use at the time.

Amador sent word to Governor Borica, recommending that the Spaniards go to war against the Saclans. The governor responded on July 10, ordering Amador to attack. According to Milliken:

> *At dawn on July 15, 1797, twenty armed Spanish horsemen struck the Saclan village of Jussent, probably located in the present Moraga*

or Lafayette area of central Contra Costa County. The Spaniards were surprised to find that the village had been divided into three separate house clusters. They approached the middle house cluster, which contained about fifty men and women (probably no more than twenty of whom were adult males capable of fighting). The Saclans had prepared ditches beforehand to keep the soldiers from charging through the village on horseback.[5]

The battle that followed lasted two hours. Soldiers advanced on Saclans who were barricaded in wells.

The Amador party continued north, leading thirty prisoners. Along the way, the raiders were attacked by Saclan men from other villages. At dawn on July 17, the party entered a Huchiun village at San Pablo Bay.

Having collected eighty-three Christians and nine non-Christians captured from the Saclan and Huchiun villages, the Amador party marched along the bay shore for three days, returning to Mission San José on July 18. Five days later, the party began a three-day walk up the peninsula to San Francisco.

The captured non-Christian Saclan and Huchiun men were put on trial on August 9, 1791. All were found guilty of various charges.

Potroy, of the village of Jussent, was implicated as the leader of the April 29, 1795 Chimenes armed defense. Potroy acknowledged that he had helped to kill the seven mission men at Chimenes Rancheria. He was sentenced to seventy-five lashes (twenty-five each on three occasions), plus one year in shackles. Caguas and Ojyugma, confederates of Potroy, were similarly sentenced. Milliken concluded:

The Chimenes battle was a catalyst for the mass flight of the population of frightened new neophytes. It also ignited a serious threat to Spanish hegemony over the many Bay Area tribes....Not a single tribal couple appeared at Mission San Francisco for baptism from the day the killings were reported until March 1800. No similar mass flight or ongoing resistance occurred again in west Central California until the late 1820s.[6]

The End of the Mission System

Despite the militancy of the Saclan resistance leaders, most Native people were ultimately forced into the mission system. There, under crowded conditions, malnourished and depressed, more than half of the population died. Within each mission, each year seemed to bring new outbreaks of disease—pneumonia, measles, smallpox, typhoid, diphtheria, scarlet fever, tuberculosis and syphilis.

By the 1820s, the mission system was in decline, despite the church having assumed control of at least one-third of the most valuable coastal land in Alta California.

In 1821, Mexico declared its independence from Spain. Soldiers who had served the Spanish Crown, as well as the *Californio* ranchers and farmers, increasingly objected to the accumulation of wealth and land by the church. The process of secularization began in 1834, with the Mexican government seizing control of mission property and sending the Franciscan missionaries into exile.

By 1836, when Mission San José was secularized, the process was complete.

Bay Area Natives Today

In the twenty-first century, many local residents with Saclan and Ohlone heritage are descendants of the survivors of the brutal mission system. Native individuals today may have genetic heritage from several different tribes.

Even among people who have similar tribal affiliations, a wide range of political views and social perspectives are voiced. But there is one matter on which there is general agreement: Natives today want the public to understand that they are alive and well.

The Muwekma Ohlone Tribe of the San Francisco Bay Area is a well-known association of local Natives that conducts historical research and interfaces with government and public agencies. According to the tribal website (http://muwekma.org):

> *The present-day Muwekma Ohlone Tribe is comprised of all of the known surviving American Indian lineages aboriginal to the San Francisco Bay region who trace their ancestry through the Missions Dolores, Santa Clara, and San Jose; and who were also members of the historic Federally Recognized Verona Band of Alameda County.*

Today, East Bay Natives form a community with ongoing social and cultural expression. Ancient traditions are being taught to young people.

Cultural groups are affiliated with local colleges and universities, as well as historical centers at both Mission San José in Fremont and Mission Dolores in San Francisco. Native voices are being heard regarding social, academic and land use policy development.

Chapter 2

CALIFORNIOS

The country…was an interminable grain field; mile upon mile, and acre after acre, wild oats grew in marvelous profusion, in many places to a prodigious height…and shoulder high with the equestrian, wild flowers of every prismatic shade charmed the eye…

The almost boundless range was intersected throughout with divergent trails, whereby the traveler moved from point to point, progress being, as it were, in darkness on account of the height of the oats on either side, and rendered dangerous in the valleys by the bands of untamed cattle sprung from the stock introduced by the mission fathers.[7]

—*M.M. Wood,* History of Alameda County, *1883*

The *Californio* Era (1769–1848)

Amid great expanses of rolling plains and hills, a small population of Spanish-speaking families created the *Californio* culture. Historian Herbert Howe Bancroft in 1888 named the years from 1769 to 1854 "California's Pastoral Era.[8]

Bound together by a provincial kinship network, the *Californio* culture made an indelible mark on the state. Brief though the years of Spanish and Mexican rule may have been, their impact lives on in California place names, styles of architecture and subtleties of legal principles. It is claimed in some contemporary circles that rivalry between the *norteño* and *susereño ranchero* families of the nineteenth century began the northern versus southern California disputes that persist today.[9]

The East Bay redwood forest extends across the boundaries of the *ranchos* (ranches) once owned by Antonio Maria Peralta, Joaquín Trinidad Moraga and Guillermo Castro. These families, their *vaqueros* (cowboys) and their neighbors played an important role in the history of the East Bay hills.

The Coming of the Spaniards

The first chronicled appearance of Spanish explorers in today's East Bay was in late 1769, when a scouting party from the expedition of Gaspar de Portolá arrived along the bay shore. Having survived the scurvy and malnutrition that killed most of their comrades on the sea voyage and overland march between mainland Mexico and San Diego, a small advance group continued on. The leaders were Lieutenant Pedro Fages and Sergeant Francisco Ortega, with their best troops from the Catalonian volunteer dragoons. Franciscan missionary Juan Crespí was the appointed diarist, having proven to be a hardier fellow than his superior, the fifty-five-year-old Junípero Serra.

The men traveled by mule train from San Diego. They made their way north along the path that would become *El Camino Real*, largely along the route of today's Highway 101.

Members of the Fages/Crespí party first observed San Francisco Bay while surveying for the Spanish government. In search of Monterey Bay, which had been described by sea captains, Sergeant Ortega led scouts around the southern end of San Francisco Bay, continuing up the east shore as far as today's Hayward before turning back to rejoin the main party. In the vicinity of Santa Cruz, they saw and named the *palos colorados*, the coastal redwoods.

On their southbound trek in early 1770, near today's Salinas, the Spaniards were greeted by people from the Rumsen Ohlone tribe. The Natives provided *pinole* (a spiced pudding made from cornmeal and seeds) for the hungry troops and were presented with glass beads in exchange.

Upon returning to San Diego, Pedro Fages was promoted to the rank of captain. He led another march northward in 1772, passing through present-day Concord, Pittsburg, San Ramon and the vicinity. Encounters with Huchiun and Carquin people were reported to be friendly.

While initial encounters between the Spaniards and the Natives were reported to involve only pleasant exchanges of gifts, subsequent explorers reported a less friendly atmosphere. Native women did not come forward to greet the Europeans.

Although Friar Crespí's diary is silent on the subject of the sexual abuse of Native women by the Spanish troops, there was clearly cause for concern. The next expedition was designed so that wives and children accompanied the soldiers. While the military sought to secure territory for Spain, the objective of the church men was to persuade the Natives to adopt the Catholic faith and the European agricultural way of life. From his post in Mexico City, Antonio Maria Bucareli, the viceroy of New Spain, reasoned that antagonizing the Natives would be counterproductive.

Colonization of Alta California

Spanish colonization of Alta California began in earnest in 1775. Lieutenant Colonel Juan Bautista de Anza and Sergeant José Joaquín Moraga led a party of about three hundred colonists, with one thousand head of livestock, from Tubac, Arizona, to Monterey, the provincial capital of Alta California. The fortitude of these people was remarkable indeed. Men, women and children of Native (*indio*), African, Latino and European descent completed the overland march. Other than Anza himself, few were older than twenty-seven.

Having brought the colonists to Monterey, and thus fulfilling their obligation to the viceroy, Anza and Moraga continued with an advance party of twelve men, including the Franciscan missionary Pedro Font as diarist. Their objective was to retrace the route of the Portolá expedition and thereby to find San Francisco Bay.

In this, they succeeded. The scouting party rode northward to the tip of the San Francisco peninsula. There, on March 28, 1776, Anza selected a location for the *presidio* (military fort). Later, after Moraga had been left in charge, a location for Mission Dolores was selected. In keeping with the desire of the missionaries to attract Native peoples into the Christian fold, the site was a comfortable distance away from the *presidio*.

Anza, Moraga, Font and their companions made a reconnaissance of the Bay Area. They spent a night on what is now the Mills College campus. On April 1, 1776, both Anza and Font reported having seen the inland redwood forest during their travels along the East Bay shore.[10] Font included the *palos colorados* in the map he drew, showing San Leandro Bay and Alameda.[11]

They continued northward to the Sacramento/San Joaquin river delta and inland into Livermore Valley. In their encounters with Natives, the

Franciscan diarist reported that the people in the Yrgin homelands in the vicinity of the redwood region evinced fear, while those in the Chupkin area farther north, near present-day Martinez, showed anger at the sight of the Spaniards.

They next encountered thirty Costanoan Natives at Alameda Creek. They extended their hands to indicate that the soldiers should stop and called out "*Au, au*!" while vigorously slapping their hands against their thighs, according to Font.

Around the same time, navigator José de Canizares drew the first accurate map of San Francisco Bay and the surrounding area. Canizares, first pilot of the packet *San Carlos*, with his captain, Lieutenant Juan Manuel de Ayala, anchored off Angel Island in August 1775. The two spent forty-four days exploring the bay in the ship's longboat. Canizares worked on his map for several years. The ultimate draft shows the East Bay redwood forest in its correct location.

The Land Grant Aristocracy

Initially, land grants were given by the Spanish monarchy to military officers as payment for their service in Alta California. Deeds were officially written by the governor in Mexico City.

Each grant of land came with a *diseño*, or hand-drawn map. Topographical features referred not only to fixed locations, such as hilltops, but also to changeable features such as willow groves and stream courses. These details led to future ambiguities that would provide income for lawyers in the years to come.

Cavalry officer Don Luís Peralta received one of the largest land grants—nearly forty-five thousand acres—in 1820. It encompassed all of the west-facing East Bay, from the ridge tops to the bay shore, from San Pablo Creek in the north to San Leandro Creek in the south.

Secularization of the Missions

Even after Mexico became an independent republic in 1821, private citizens objected to the concentration of wealth that the mission system had accumulated. By 1830, the Catholic Church owned one-third of the land acreage in Alta California. Provincial governor José Figueroa issued a proclamation in 1834 that the missions were to be secularized. The Mexican government confiscated land and property from the church. Spanish-born clerics were sometimes exiled as "illegal immigrants."

In theory, half of the mission land was supposed to be returned to the Natives, whose labor had enriched the *padres*. Through unscrupulous dealing, however, most land and livestock fell into the hands of the *ranchero* families. Upon leaving the mission systems, Native men went to work without pay on *ranchos* as *vaqueros*; women became domestic servants. Some Natives dispersed to the Sierra Foothills to join other tribes.

Despite secularization, the Catholic Church still managed to benefit economically from the *ranchos*. According to the nineteenth-century historian William Heath Davis:

> *The Missions extracted from the cattle owners a contribution known as a* diezmo, *for the support and benefit of the clergy and for the expense of the Missions—one-tenth of the increase of the cattle. The tax was not imposed by the general government, but was solely an ecclesiastical matter decreed by the Pope of Rome or a law of the church, diligently collected by the clergy of the different Missions and religiously contributed by the* rancheros. *The collection was continued as late as 1851 or 1852.*[12]

Life on the *Ranchos*

Of all the descriptions of the *Californio* way of life, one of the most comprehensive is the lengthy memoir *Sixty Years in California*, published in 1889 by William Heath Davis. The Hawaiian-born Davis was the son of Boston sea captain William Heath Davis Sr. and Hannah Holmes Davis. His materal grandmother was Mahi Kalanihooulumokuikekai, a chief from Maui. His mother was also related to Supreme Court justice Oliver Wendell Holmes. Davis spent most of his life as a merchant in California. He married Maria de Jesus Estudillo.

Davis noted that the *Californios* were early risers:

> *The ranchero would frequently receive a cup of coffee or chocolate in bed, from the hands of a servant, and on getting up immediately order one of the vaqueros to bring him a* [horse]*....He then mounted and rode off about*

> *the rancho, attended by a vaquero, coming back to breakfast between eight and nine o'clock.*
>
> *This breakfast would be a solid meal, consisting of* carne asada… *with onions, eggs, beans,* tortillas, *sometimes bread and coffee, the latter often made of peas.* [After breakfast] *he would ride off again around the farm or to visit the neighbors. He was gone till twelve or one o'clock, when he returned for dinner, which was similar to breakfast, after which he again departed, returning about dusk in the evening for supper, this being mainly a repetition of the two former meals.*
>
> *Although there was so little variety in their food from one day to another, everything was cooked so well and so neatly and made so inviting, the matron of the house giving her personal attention to everything, that the meals were always relished.*[13]

Californio Women

Women on the *ranchos* spent their time on domestic crafts such as sewing and lace making, as well as church activities and gardening. In the more affluent families, much of the housework was handled by Native female servants.

Davis, who was much enamored of *Californio* women as a group, described their role in society:

> *The women were full of natural dignity and self-possession; they talked well and intelligently, and appeared to much better advantage than might have been supposed from their meager educational facilities…*
>
> *Their husbands oftentimes looked to then for advice and direction in their general business affairs…*
>
> *There were no established schools outside of the Missions, and what little education young people obtained, they picked up in the family, having learned to read and write among themselves.*[14]

Natives on the Ranchos

After their release from the mission system, Natives went to work on the *ranchos* in a feudal relationship with the wealthy landowning families. They were provided with food, clothing and shelter on the *ranchos*. According to

William Heath Davis, "Civilized Indians from the missions were scattered about the country and many were to be found on the different *ranchos.* They were of peaceable disposition, were employed as *vaqueros*, and helped the *rancheros* at the planting season and at harvest time."[15]

Labor provided by Natives included herding and slaughtering cattle, processing tallow in iron pots and shearing sheep. Natives tilled the soil for *rancho* gardens, as well as wheat, corn and bean fields. They harvested the crops; they ground and threshed the grain. Natives did the hard work of making adobe bricks and tiles from which they built *haciendas* (houses), workshops and churches. They paddled the boats in which *Californio* families were transported to San Francisco to shop for the manufactured goods that arrived on ships. Household services performed by Native women included care of the *rancho* children, cooking, washing dishes and laundry.

Farming and Milling

Each *rancho* had a large garden of corn, beans, squash, tomatoes, peppers and herbs. In the California climate, two crops per year of beans and corn were typical. But for tortillas and other foods, wheat farming was essential. Larger *ranchos* operated flour mills. In some cases, mill stones were ordered from Spain and carried by ship around Cape Horn.

In the town of Niles, south of the East Bay redwood forest, José de Jesús Vallejo constructed a primitive grain mill before 1852. In that year, E.L. Beard and H.G. Ellsworth built a larger mill at Mission San José. In 1853, Vallejo was inspired to build yet a larger mill at Niles.

Ranching and Trade

The primary business pursuit of the *Californio* men was ranching. The cattle grazed on oats that the Spanish farmers had planted, a grain that soon crowded out other native grasses. This diminished the availability of seeds that were once a staple of the indigenous peoples' diets.

Hides and tallow were traded to merchants and sea captains in exchange for manufactured European goods. The tallow ships were called "greasers," leading to the use of the term as an offensive epithet.

Hides were tanned and made into leather goods of all sorts—saddles, furniture and even leather breastplates that could repel an arrow.

Tallow (rendered beef fat) was the raw ingredient for candles, soap and other products. The most highly valued form of tallow was called *manteca*, which was the best for candle making. The tallow was cooked down in large iron kettles at the slaughter site. Cauldrons used for this purpose are on display in the Mission San José and Peña Adobe (Petaluma) museums.

At roundup time, the rodeo would be a festive event to which the denizens of neighboring ranches were invited. *Vaqueros* came along so that if cattle bearing the brand of their ranch had comingled with the host's cattle, the animals could be cut out of the herd and driven home.

William Health Davis wrote:

> *The cattle were slaughtered in the summer season; the killing commenced about the first of July and continued until the first of October for the hides and tallow; about 200 pounds of the best part of the bullock was preserved, by drying, for future consumption, the balance of the animal being left to go to waste; it was consumed by the buzzards and wild beasts.*
>
> *The tallow was "tried" in large pots brought to the American whale ships—such as are used to try out their blubber, and was then run into bags made of hides, each containing twenty to forty* arrobas. *An* arroba *is twenty-five pounds.*

Grizzly Bears

As long as the elk population remained stable, the *Californios*, and later the Yankee men, slaughtered them in the same wanton fashion as the bison of the Great Plains. Given the practice of leaving skinned carcasses of cattle and elk on the open range, grizzly bears naturally came to partake of the meat. Bears were lassoed by *vaqueros* for fun and excitement—and as a show of skill and *machismo*. On occasion, bear and bull fights were staged for a bloodthirsty diversion.

Davis described a bear encounter he experienced on the East Bay shore sometime in the 1830s:

> *At one time I was encamped at the embarcadero of Temescal, a place between where the Oakland long wharf and Berkeley are now, in order to receive hides and tallow from the cattle that were slaughtered not far away....One night I sent my man up to Don Vicente Peralta's house of an errand, and remained in my tent alone all night, to my great peril, as I soon discovered.*

> *...Peralta and his vaqueros came down in the night to lasso the bears for sport. Some of them got away from their enemies and made for my tent....I sat in the tent and heard these animals circling round and round for several hours, going off at times and returning. I was in constant fear that they might...devour me...*
>
> *On giving Don Vicente Peralta a narrative of my narrow escape from being devoured by bears which he and his vaqueros had stampeded to my tent, he laughed heartily, but became serious when he realized the gravity of my situation....After this occurrence, whenever I had occasion to stop over night there, he would send a vaquero with a horse, and kind messages from himself and wife to be their guest for the night, which invitations I gladly accepted.*[16]

Horsemanship and Other Entertainment

Californio women of distinguished families conducted two types of social events. For the *baile* (ball), invitations were carefully issued to select guests for elaborate dancing parties. In contrast, the *fandango* was a festive gathering at which the entire community was welcome. Regardless of the social stratum, parties could go on for a week or more. Many days were spent with both women and men riding and enjoying picnics. Nights were spent on the dance floor.

William Heath Davis noted that many of the young women played the guitar, and many young men played the violin.

Each Sunday, the young ladies and gentlemen from the elite families would parade in the local plaza, showing off their skill at horsemanship. Their fine mounts were of Arabian bloodstock. Only stallions and geldings were ridden for ranch work or for show. Mares were used only for breeding and for agricultural tasks such as threshing grain.

After the roundup and ensuing *matanza*, *Californio* men participated in rodeos. Even the elderly *dons* demonstrated their skills. According to M.M. Wood in his 1883 *History of Alameda County*, "When not sleeping, eating, or dancing, the men passed most of their time in the saddle and naturally were very expert equestrians. Horse-racing was with them a daily occurrence, not for the gain which it might bring, but for the amusement to be derived therefrom; and to throw a dollar upon the ground, ride at full gallop and pick it up, was a feat that almost any of them could perform."[17]

Davis described the horse races and bear or bull fights that enlivened celebrations of saints' feast days:

> *The vaqueros were…relieved from duty, wore their best clothes, and were allowed to mount their best horses and to have sport.…Bets were made in cattle and horses…at times one hundred up to several hundred head of cattle were bet on the result of a single short race…they had no money to wager, but plenty of cattle.*
>
> *The bull was turned into an enclosure, and the horsemen would come in, mounted on their best animals, and fight the bull for the entertainment of the spectators, killing him finally.…Sometimes a bear and bull fight would take place, another amusement that they had at the killing season at the* matanza *spot. When the cattle were slaughtered, the bears came to the place at night to feast on the meat that was left.…* [T]*he rancheros with vaqueros would go there for the purpose of lassoing them. This was one of their greatest sports; highly exciting and dangerous, but the bear always got the worst of it.*[18]

The Casta System

People of the *Californio* culture were highly conscious of racial heritage. Initially, many of the Spaniards who came to the New World were, for their time, highly educated men. Spanish-born military officers and Catholic priests typically had the benefit of a classical education. But the Spaniards' sons who were born in the New World did not have access to the same level of education.

Many *Californio* gentlemen of the nineteenth century, and most of the ladies, were illiterate. Cecelia Peña noted, however, that some families, including her own, valued education highly for both women and men.[19]

Deriving from this military and church hierarchy, the *casta* system evolved. Social status was assigned according to how much Spanish blood an individual supposedly had inherited. According to historian Gloria E. Miranda, "The Spanish self-identity fascination of early California was the direct by-product of two and one-half centuries of Spain's colonial racial and social stratification policies."[20]

The Moraga Land Grant in the Redwoods

It was Joaquín Trinidad Moraga, along with his cousin Juan Bernal, who in 1835 was granted the east-facing share of the inland redwood forest, later to be known as the Moraga Redwoods.

Joaquín Trinidad Moraga was the grandson of José Joaquín Moraga, who had been Anza's second in command. Bernal was also descended from a cavalry officer who had come to California with the Anza party.

Moraga named his spread *Rancho Laguna de los Palos Colorados* ("Ranch of the Lake of the Redwoods"). The grant encompassed much of the acreage that is present-day Lafayette, Moraga and Orinda, plus part of the districts of Canyon and Rheem. The eponymous lake no longer exists. In wetter times, before the area was developed and the groundwater was depleted, the lake was near today's Campolindo High School campus.

Moraga built his adobe house on an Orinda hillside around 1841. This is one of the few homes from the *Californio* era that has survived into the twenty-first century.

Diarist Joseph Lamson in January 1855 described his experience of hiking over the hills from his general store, located near the entrance to today's Redwood Regional Park, to attend a festive *fandango* at the Moraga homestead:

> *I received an invitation to attend a party at* [the Moraga adobe] *on New Year's eve, 1855....His house overlooked a beautiful valley...*
>
> *Señor* [Joaquín] *Moraga, the father of our host and owner of the estate, an old man of seventy, short, thick, corpulent and coarse-featured, but sprightly, active, and polite. Then his sons, José and Francisco, between thirty and forty years of age, swarthy men with very good features, black hair, whiskers and mustaches. They were very gentlemanly in their deportment...*
>
> [F]*oremost among* [the ladies] *was Doña Maria, our hostess, and the lady of José Moraga....Her black, glossy hair was arranged in the usual Spanish style, in two braids that hung down her back. She was dressed in a black silk that fitted well her capacious person. She had several daughters... who were very pretty dancers....I observed a considerable number of Indian women in the house, and there was no lack of papooses among them...*
>
> *Two musicians had been employed for the occasion. Their instruments were a violin and a guitar. Dancing was the principal amusement...*
>
> *One of the gentlemen arose and waved his handkerchief toward a lady, whereupon she arose...one of the gentlemen seized his neighbor's hat...and placed it upon the young lady's head. She still continued to dance without paying the slightest attention to this apparently uncivil act...another and another of the ladies were...motioned up, who each performed the same dance, and each was similarly crowned with a hat or handkerchief, and sometimes with several of each....All this was carried on with great*

> *merriment on the part of the young fellows....* [My] *own* sombrero *was suddenly snatched from my head, and placed on that of a young señorita....I was then informed that...* [it] *must be redeemed by a payment of* [half a dollar]...*to the fair one...*
>
> *Supper was ready at an early hour...soup, baked meats, boiled chickens and bread, with wine in glass tumblers...*
>
> *Thus pleasantly passed the evening until eleven o'clock, when...I bade them* á Dios, *and wended my way back over the mountains to my lodgings. The company continued dancing till morning.*[21]

Juan Bernal, Moraga's cousin and land partner, died in 1847, before the Treaty of Guadalupe Hidalgo ceded Alta California to the United States. Thereafter, in 1853, Moraga filed a claim for *Rancho Laguna de los Palos Colorados* with the Public Land Commission. Before the court confirmation could be completed, Moraga himself died in 1855. The grant was finally patented to his heirs in 1878.

By then, however, lawyer Horace Carpentier had acquired most of the property through a complex series of legal maneuvers—in a similar fashion to how he acquired land from the Peralta family grant on the west side of the hills.

The Peralta Family

Don Luís María Peralta came with his parents and siblings to California from his native home in Tubec, Sonora, Mexico, in 1776 at age seventeen, serving as a military escort for the Anza expedition.

In 1820, as compensation for his forty years of military service, Don Luís Peralta received the land grant comprising much of today's East Bay. The signatory was Don Pablo Vicente de Solá, the last Spanish governor of California. This five leagues of land extended from "the deep creek of San Leandro" in the south to Albany hill (*el cerrito*) in the north and from the ridge line encompassing Redwood Peak and Grizzly Peak in the east to the San Francisco Bay shore.

The text of the Peralta land grant, written by Father Narcisio Duran of Mission San José, contained a clause dictating that "cutting of wood must remain in common." In other words, Peralta did not own the redwood timber and could not harvest the trees for commercial gain. However, since the government retained the timber rights, Peralta was given the authority

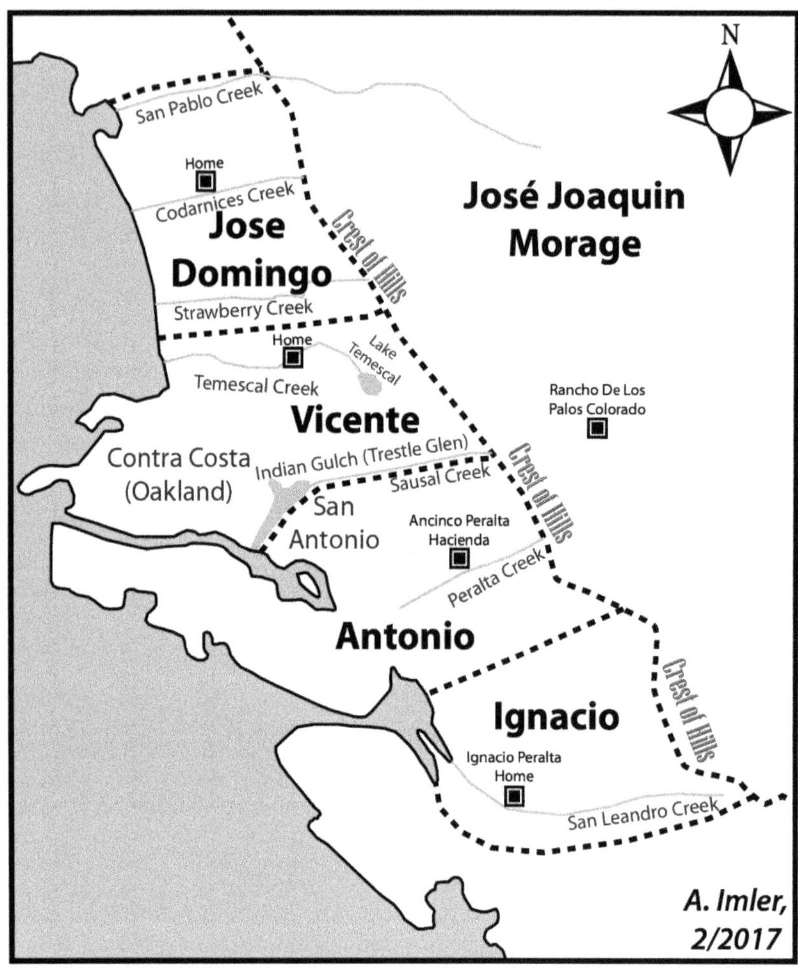

The four Peralta brothers, sons of Don Luís Peralta, inherited segments of the East Bay coastal plains between San Pablo Creek in the north and San Leandro Creek in the south. *Courtesy Alan Imler Cartography.*

to summon the military to repel illegal loggers—a right he would eventually need to exercise.

Don Luís named his East Bay land Rancho San Antonio. The village at the wharf where the redwood logs were loaded onto ships—at the foot of today's Fourteenth Avenue in Oakland—was also named San Antonio.

Don Luís, however, was sixty-two years old in 1820 and well settled in at his *rancho* near Mission Santa Clara with his unmarried daughters Josefa and Guadalupe. So, Don Luís promptly passed the East Bay land on to his four sons. At first, the sons were given joint use of the property, but in August 1842, Don Luís formalized the division into four parts. The divisions were made using major creeks or imaginary lines drawn from the ridge top to the bay. In 1851, Don Luís formally divided the acreage, all the better to press the family case in the United States courts.

To occupy his share of the land grant, the eldest son, Ignacio, retired from his position as *alcalde* of San José. Ignacio built a home for his family on San Leandro Creek. His land extended north as far as today's Sixty-Eighth Avenue in East Oakland and encompassed today's San Leandro and San Lorenzo. A later addition to his homestead was a brick house built in 1860; this and the wooden house built by Antonio near today's Coolidge Avenue in Oakland are the two Peralta houses still standing today.

The second-born son, Domingo, had settled on his own *rancho* in Santa Clara at the time the land was divided. His portion of the land grant extended

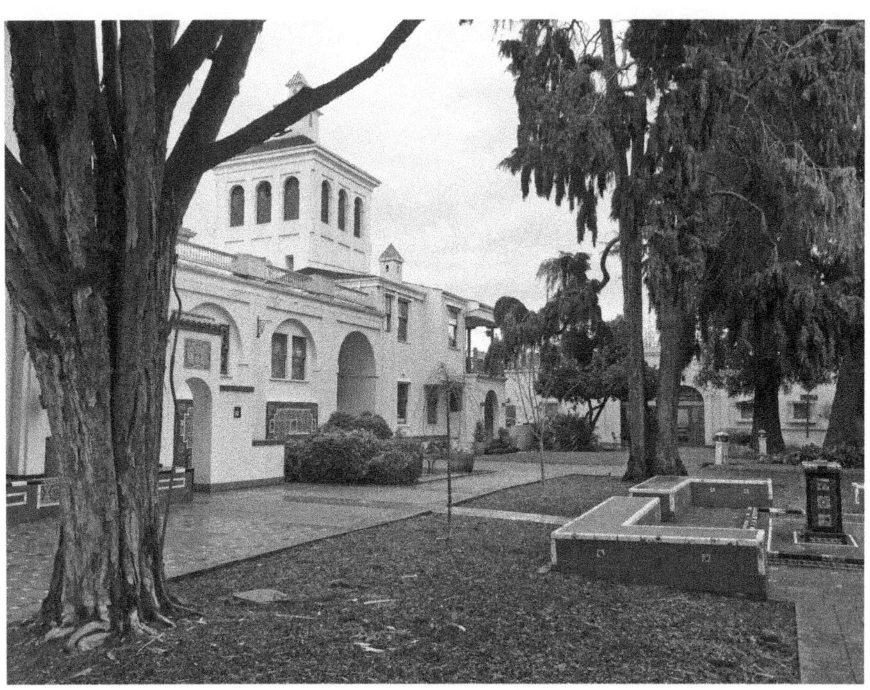

Heirs of Ignacio Peralta built this elegant *hacienda* on San Leandro Creek. Casa Peralta Museum is managed as part of the San Leandro Public Library. *Amelia S. Marshall photograph.*

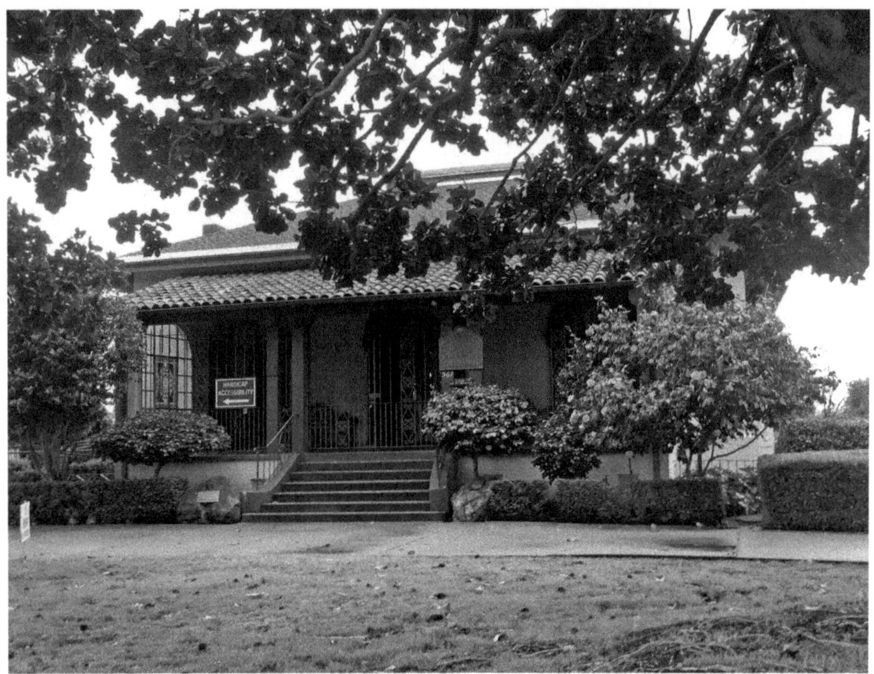

Ignacio Peralta owned the first brick house in Alameda County. The landmarked Peralta Home has been the Alta Mira Club since 1926. *Amelia S. Marshall photograph.*

from Albany hill to Temescal Creek, including all of today's Albany, Berkeley and parts of North Oakland. After the secularization of the missions, in 1836, he moved with his family to his father's land grant. In 1841, he built an adobe house for his family on the south bank of Codornices Creek, near today's 1304 Albina Avenue in Berkeley. (The creek itself was named by Domingo for the California quail. According to legend, on the day Domingo and one of his brothers first discovered the creek, they were hungry. They found a quail nest there on the bank and ate the eggs.)[22]

The youngest son, Vicente, lived with the third son, Antonio, at the *rancho* on Peralta Creek until he married in 1836. Then he built an adobe house on the spectacularly beautiful Temescal Creek, marked by a grove of oaks called Encinal de Temescal.

William Heath Davis described Vicente Peralta, with whom he was acquainted through business and kinship:

> *Don Vicente was about six feet tall, finely proportioned, straight as an arrow, weighing about 225 pounds, hospitable, kind, and full of native*

> *dignity. His surroundings were in keeping with his appearance, manners, and taste…*
>
> *On one occasion in 1840, I stopped at his house during one of my trading expeditions, remaining over night. In the morning he said to me, "Let's take a ride out this beautiful April morning. You see how handsome the hills are; it is the pleasantest part of the year. Just now the cattle and horses are beginning to change their coats and everything is fresh and new."*[23]

The site of Vicente's adobe is now on the east side of Telegraph Avenue, south of the Highway 24 overpass. His portion of land extended south to Lake Merritt.

It was Antonio Peralta whose land portion was the largest and included the redwood forest. The southern boundary adjoined Ignacio's land at Sixth-Eighth Avenue in Oakland and extended north to Lake Merritt, including the then-peninsula of Alameda. Antonio also received the land on the eastern side of Lake Merritt extending up to Indian Gulch—today's

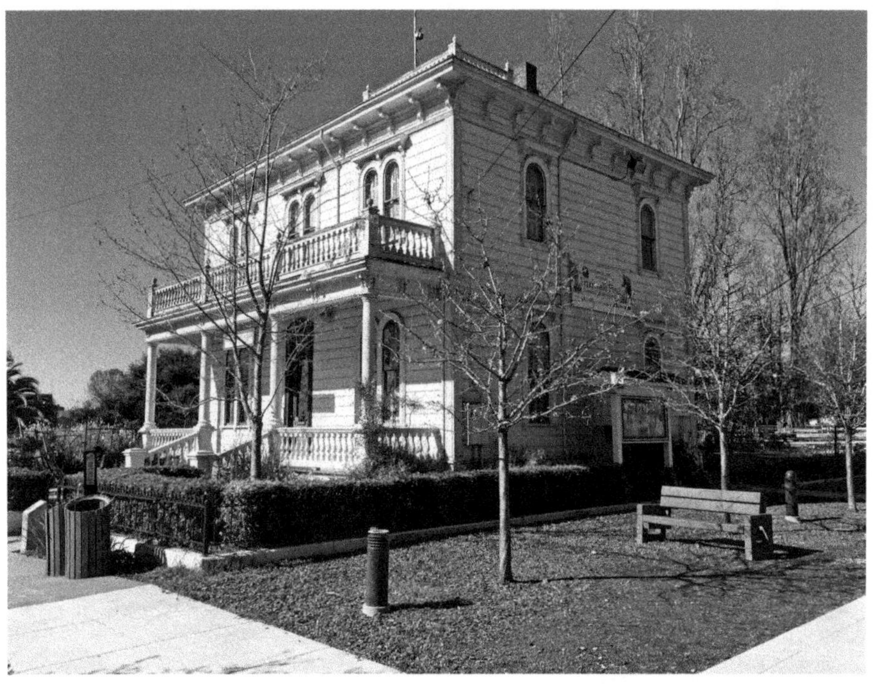

Peralta Hacienda Historical Park and Museum, Antonio Peralta's home in Oakland, provides community and educational programs, as well as museum exhibits and archives. *Amelia S. Marshall photograph.*

Trestle Glen district. The land grant extended up to the hilltops, where it adjoined the land grant of the Moraga and Bernal families.

Antonio began work on the first adobe building for his family *hacienda* on Peralta Creek in 1821. This was the first non-Native habitation constructed in today's East Oakland.

Don Luís remained at his Santa Clara *rancho* until his death in 1851, when he was ninety-two years old.[24]

Antonio Peralta and His Household

The first wife of Antonio was María Antonia Galindo. They had eleven children. María died in 1850. No photographs of her have ever been discovered, and she left no journals or correspondence.

The widower Antonio thereafter married María Dolores Archuleta, who lived until June 1868. In October of that year, an earthquake on the Hayward Fault badly damaged the 1821 adobe house. The larger 1840 adobe house was destroyed along with twenty-two lean-to shacks. Thereafter, Antonio built

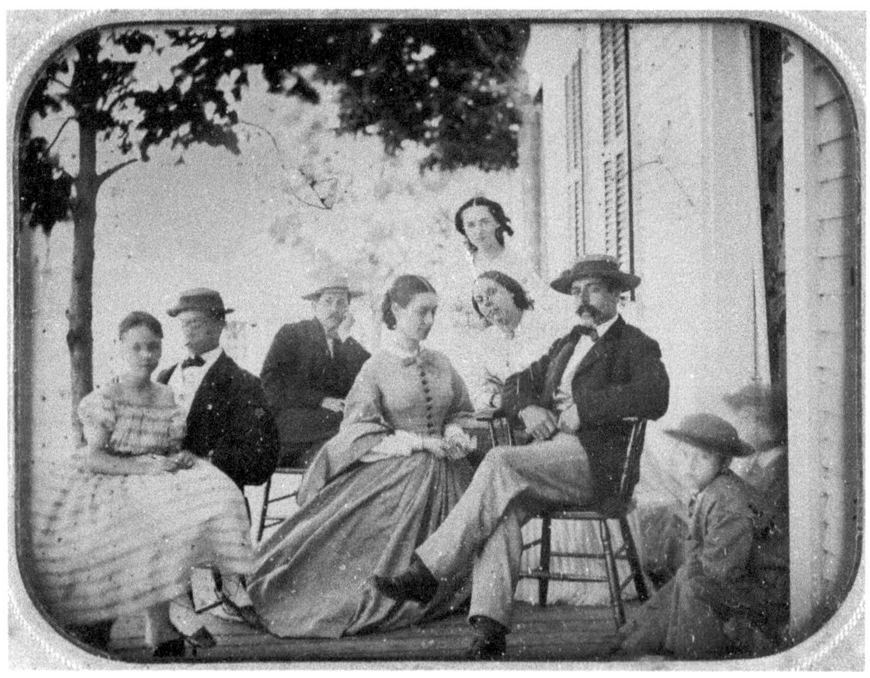

Antonio Peralta and his household are shown at leisure. Social activities played an important role in *Californio* society. *Courtesy Bancroft Library.*

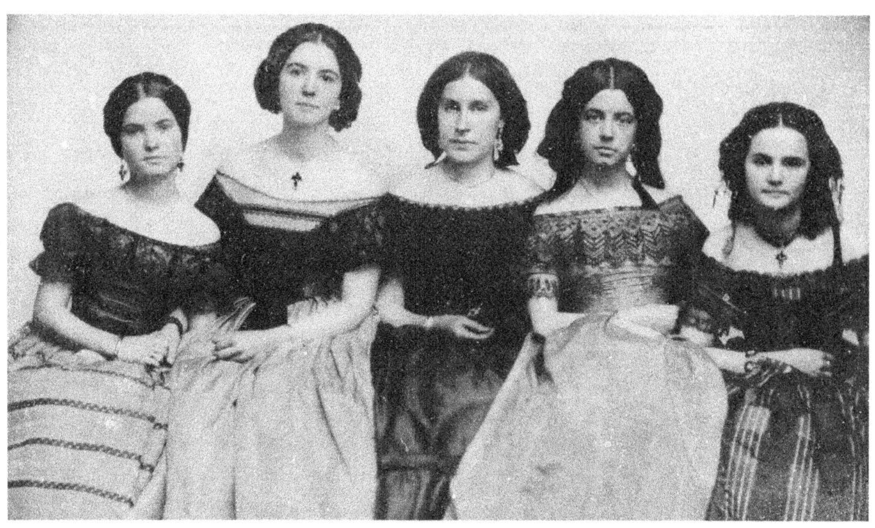

These ladies of the Antonio Peralta household are tentatively identified by the Peralta Hacienda Museum as Ynez and Rosa Peralta, María Dolores Archuleta, Rosa Valencia and Paula Peralta. *Courtesy Bancroft Library.*

the wooden Italianate farmhouse that now serves as a museum in the Peralta Hacienda Historic Park. He died in 1879 at the age of seventy-seven.

Antonio bequeathed the house and twenty-three acres of land that remained of his once-extensive *rancho* to Francisco Galindo, husband of his youngest surviving daughter, Ynez, to repay a debt.

In 1897, Ynez Peralta Galindo sold the Peralta House and surrounding land to real estate developer Henry Z. Jones, who subdivided the property into city lots. Jones demolished the last wall standing from the original 1821 adobe *hacienda*. The adobe bricks were sent to Oakland's Dimond Park, where they were used to construct a Boy Scout lodge. The structure, with its antique adobe bricks, remains in Dimond Park today.

YANKEE ENCROACHMENT

By 1840, the San Francisco Bay Area had become a global destination. Wanderers, fortune seekers and disenchanted sailors came from afar, including Yankees, English, Portuguese, French and countless others. The most obvious source of timber was the redwood forest. Soon illicit logging had begun.

William Heath Davis arrived from his native Hawaii in 1838. Eugène Duflot de Mofras was sent as an emissary to the Department of California by the French government in 1839. He served as a diplomatic attaché, naturalist, cartographer and foreign correspondent. Naturally, he and Davis became drinking buddies.

Mofras wrote in his diary in 1841 that French naval deserters Sicard and Leroy were logging the redwoods. He further noted that "[s]treams that descend from the mountains are navigable by small boats that take on loads of planks and timber."[25]

Antonio Peralta Summons the Cavalry to Repel Sutter's Loggers

William M. Mendenhall came west from Ohio to seek his fortune. In 1845, when he was twenty-two, he arrived at New Helvetia,[26] on the Sacramento River, the fort and colony that had been established by merchant John A. Sutter. Accompanied by two brothers, Henry Clay Smith and Napoleon Bonapart Smith, Mendenhall sailed Sutter's launch downriver to the town of Yerba Buena (San Francisco).

M.W. Wood described the first encounters in 1846 between the newcomers and members of the Antonio Peralta household in Yerba Buena, today's San Francisco:

> [T]*hrough the agency of the British Consul, they crossed the bay to where the city of Oakland now stands, and from thence started on foot for the San Antonio redwoods, but meeting two Spaniards, their passports were demanded, which not being able to produce they were frightened away with the fire-arms of the party.*
>
> *They ultimately got to the redwoods, however, and there engaged with a Frenchman for some time in making shingles and sawing lumber, but this not meeting the views of the irate Dons, twenty soldiers were brought into the redwoods to oust them. Mendenhall and his party, however, believing discretion to be the better part of valor, secured what horses they could, with their effects, and beat a hasty retreat into the valley of the San Joaquin.*[27]

According to the terms of the original land grant to Don Luís Peralta, while he owned the land, he did not own commercial timber rights to the redwood forest. The Spanish monarchy had intended to keep this prime resource, with an eye toward potential ship building.

In 1846, Don Antonio called in the cavalry to protect the redwood trees. Accordingly, the twenty soldiers who drove out Mendenhall and the Smith brothers came from the presidio of San Jose in response to Antonio's summons.

Environmental Change

Rancho life and the cattle industry in California were greatly affected by bad weather in the 1860s. Floods devastated the cattle ranges in 1860 and 1861, followed by the "Great Drought." Soon the open ranchland was fenced into smaller lots, accommodating an increased population. With this loss of habitat and no more skinned cattle carcasses for food, as well as hunting by humans, the grizzly bear population crashed. The last grizzly bear in the East Bay was seen in a San Leandro orchard in 1863.

The Peraltas Lose Their Land

With the annexation of California to the United States, the expansive Peralta landholdings were gradually diminished. Between 1849 and 1853, the influx of English-speaking settlers quadrupled the population of the East Bay. To the eye of the Yankee newcomers, the unfenced open rangeland did not appear to be private property. The United States Preemption Act of 1841 allowed settlers to claim public land. Moreover, many Yankee squatters took the attitude that the Mexicans had lost the war and, therefore, the *Californios* had no land rights.

Squatters

In 1853, the squatters formed an association for collective bargaining and mutual defense called the Pre-emptioner's League (*sic*); 109 men signed a proclamation and paid five dollars. In part, their manifesto read:

> *We the undersigned, citizens of Alameda County, and settlers upon what are supposed to be the public lands belonging to the United States, within said county, believing we can more effectually guard our interests as such*

> *settlers by mutually supporting and protecting each other: Therefore form ourselves into an association....* [I]*ndividually we will make no overtures to the land claimants for a settlement of our difficulties with them and will reject all such as may be made to us by them until such overtures shall have been submitted to and approved by this League....* [A]*t all times* [we] *will hold ourselves in readiness to aid and assist each other to defend our homes and farms from the grasping avarice of land speculators.*[28]

A county convention of this league was held at the *rancho* of W.R. Richardson, land that had been part of Antonio Peralta's grant, on October 29, 1853. Alameda County historian Wood quoted from a record book kept on this occasion:

> *The sun shone gloriously, as if heaven smiled on our cause, and the old cannon, "The Squatter," belched forth its thunders, calling together the farmers around. On every side could be seen the hardy pedestrian and horseman, and four-horse teams with the "Stars and Stripes" floating gaily in the breeze.*[29]

Diarist Joseph Lamson, who operated the general store in the Middle Redwoods, described a conversation in 1857 with Antonio Peralta concerning the Yankee interlopers:

> [A] *great part of* [Don Antonio Peralta's] *lands have been seized by squatters by virtue of an act of the legislature giving to each settler a preemption claim to a quarter section, or one hundred sixty acres of land, belonging to the State. Peralta's lands, as well as those of nearly every other Mexican landholder in the State, have been taken in this manner, in the hope that his claim would be rejected by the United States Commissioner appointed to adjudicate land titles, when the squatters would become owners of valuable farms at a merely nominal price...*
>
> *I was at Peralta's house some months since, when a man in the course of conversation used the word "squatter." Peralta could not, or would not, speak English, but he well understood the significance of that word, and I was struck with the expression of scorn and hatred which he threw into his countenance as he slowly repeated, "Squatter! Squatter!"*[30]

Property Taxes and Lawyers

Another development new to the *rancheros* was the institution of property taxes. The cash-poor *Californios* were forced to sell land in order to pay these taxes. *Rancheros* were required by the United States Congress to prove their landownership in court. Paying legal fees also led to lands being sold off. The average lawsuit took many years to resolve. The hand-drawn *diseño* maps identified property boundary lines using landmarks such as large trees, rocks and stream courses—all features that were easily altered.

Following the death of Don Luís, the Peralta family continued with litigation in the United States courts to confirm, or "patent," their title to the land. In the course of fighting the squatters and subsequent owners who had bought Peralta land from speculators, an internal family dispute erupted. The four surviving daughters of Don Luís Peralta decided to assert their own inheritance rights.

According to an article entitled "1500 Squatters" in the *Daily Alta California* newspaper dated February 11, 1854, a lawyer named Simpson made a motion on behalf of the Peralta sisters to add them as claimants to the Peralta land, along with their four brothers.

In perhaps the earliest exercise of community property law inherited from the Spanish legal system, the Peralta sisters argued that their late mother, María Loreto Alviso, had been half-owner of the estate. While Don Luís had left a lengthy and detailed will, leaving practically everything to the sons, María had died without a will.

Through the Spanish common law principle of community property, the lands in question had belonged equally to Don Luís and Doña María. The sisters claimed that only half the land was Luís's to bequeath. By the laws of intestate succession, the sisters and the brothers should each inherit a share of their mother's half. But the state Supreme Court in 1868 did not accept this argument and dismissed the "Sisters' Title" suit.

The four Peralta brothers did succeed in getting the court to accept Don Luís's original *diseño* as proof of landownership, but the suit took seventeen years. Eventually, in 1859, the Peralta brothers' claim to the land title was affirmed by the Supreme Court. However, this victory came too late for the brothers to retain their holdings.

Domingo and Vicente Peralta

Domingo and Vicente Peralta had inherited land that was on the main road, following the approximate route of today's San Pablo Avenue, from the new town of Oakland on the way to the Sierra gold fields. Most of Domingo's and Vicente's land either came into Yankee hands through fraud or was sold by the brothers under duress. The land inherited by Antonio and Ignacio, however, was less desirable to the English-speaking real estate developers, so these two brothers were able to hold out longer.

The land belonging to Vicente Peralta appealed to Horace Carpentier, a New York lawyer and real estate speculator who arrived in the East Bay in 1848. Carpentier occupied the land. When Vicente objected, Carpentier agreed to a lease arrangement. Next, however, without having legal title to the land, Carpentier laid out the streets for a new town he named "Oakland" and sold the land to others.

Carpentier was elected the first mayor of Oakland in 1854. He got 368 votes, a number that exceeded the total population.

Domingo Peralta was brought to trial after he accosted two squatters and beat them with a sword. A jury found him guilty and fined him $700. The sheriff sold 428 acres of his land to satisfy the judgment. Domingo confided in a letter to a family member as to his troubles with Horace Carpentier:

> *Sobrino, I am sore pressed by these gringos. They take my lands. First a piece here, then a piece there. I take my case to the court…the jury, which is composed of squatters, decides against me. And now this* caballero, *Don Horacio…* [h]*e comes riding on his white horse, you have seen him, have you not, wearing his blue cape with the brass button, his blue pants and large black* sombrero*? He is very fair of skin, has blue eyes, is not a very tall man, neither handsome nor yet ugly. He came only last week, dined and spent the night.…* [H]*e speaks such beautiful Spanish. He finally told me that if I signed some papers, written to be sure, in English, he would help me to win against the squatters, and I have signed. But I do not feel easy about it.*[31]

Finally, nothing was left of Domingo's land except for his house on Codornices Creek. In 1865, he died. Thereafter, Carpentier arrived to evict the family, Domingo's widow and their children, from their home. Abel, one of the young sons, threw hay around the parlor and led two horses around inside, as a parting insult to the new owner.

Antonio's Holdings

Antonio was forced to sign over 175 acres of Rancho San Antonio to his lawyer, Carlos B. Strode. This property had been fenced in by a squatter named W.R. Richardson. The lawyer then sued Richardson for title. Correspondence from 1856 shows that additional land disputes forced Antonio, with great shame, to borrow money from another lawyer, Diego Forbes, who was a member of the extended family.

Ignacio's Inheritance

Ignacio Peralta, unlike his brothers, was able to retain much of his land wealth. Since his family home was south of the main road from the Oakland *embarcadero* toward the gold fields, Ignacio escaped the scrutiny of speculators. Ignacio and Rafaela Sanchez Peralta continued to live in their adobe home on San Leandro Creek, across from Jose Joaquin Estudillo and his family. After W.P. Toler, the new husband of daughter Maria Antonia, built a house nearby, Ignacio asked Toler to build one for him and Rafaela as well. The resulting Peralta home, at 561 Lafayette Avenue, was the first brick structure in Alameda County.

Legacies of the *Californio* Era

The Contemporary Kinship Network: Los Californios Descendants

Present-day descendants of the original *Californio* Hispanic families have a statewide organization that holds events and maintains an online presence (http://www.loscalifornianos.org). In addition to its conferences and *fiestas*, this organization promotes the history of the Spanish-speaking era before California joined the United States. Online forums are devoted to genealogy research and other areas of interest.

Cactus Fencing Along Ridge Lines

Hikers who explore the East Bay hills will occasionally discover a surprising nonnative plant on high ridges: beavertail cactus. Sporadic examples can be found on ridges above the Butters Canyon neighborhood. An extensive thicket of beavertail cactus engulfs the hillside on McDonell Avenue in Leona Heights, just below the colorful sulfur mine. Incidentally, the street was named for Angus McDonell, the mine foreman, whose home was near the mine. The cactus appears to have been planted long ago.

According to Holly Alonso, director of the Peralta Hacienda Historical Park, there is no direct evidence that the cactus was originally planted by *vaqueros* from the Peralta *rancho.* However, she noted, cactus is widely used for cattle fencing throughout northern Mexico and the American Southwest.

Could cowboys have planted cactus as a way of marking boundaries between Peralta and Moraga land grants? Could the plants have been intended to deter cattle from straying—or to discourage grizzly bears from coming to feast after the roundups?

Chapter 3

SEA CAPTAINS

The marks when on [Blossom Rock] *are the north end of Yerba Buena Island in one with two trees (nearly the last of the straggling ones) south of Palos Colorados, a wood of pines situated on the top of the hill, over San Antonio, too conspicuous to be overlooked.*
—*Captain Frederick W. Beechey,* Narrative of a Voyage to the Pacific and Beering's Strait, *1832*

Sighting on the Redwoods

Before the urban forest of nonnative trees was planted in the early twentieth century, the East Bay hills appeared from a distance as rolling grasslands. Early in the year, the hills were green, with seasonal carpets of California poppies and other wildflowers. Thereafter the hillsides were golden, except where tall trees grew in the vicinity of Redwood Peak, 1,619 feet above sea level. Viewed on clear day from San Francisco Bay, the redwood forest was a distinctive landmark.

California coastal redwoods, *Sequoia sempervirens*, were first seen by Europeans—members of the Portolá expedition—in 1769, in the vicinity of the Pajaro River near today's Watsonville. Diarist Juan Crespí described them.[32]

It is likely that the first non-Native visitors to actually climb into the East Bay hills to see the *palos colorados* ("tall red trees") up close were members of

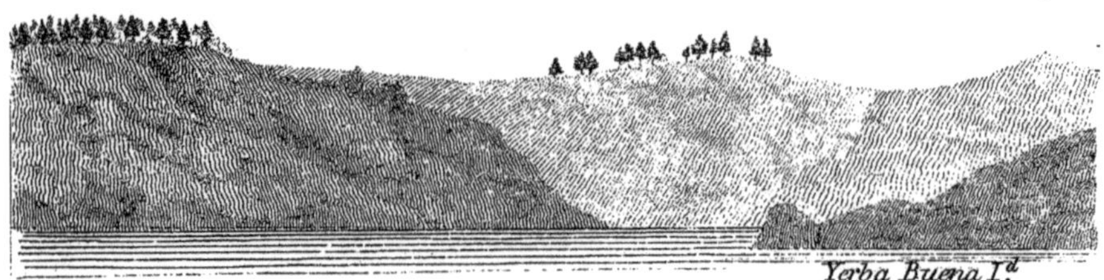

Captain Beechey's view from San Francisco Bay—giant redwoods on the hilltops. This detail from his chart may have been redrawn in a studio based on his field sketch. *Public domain map.*

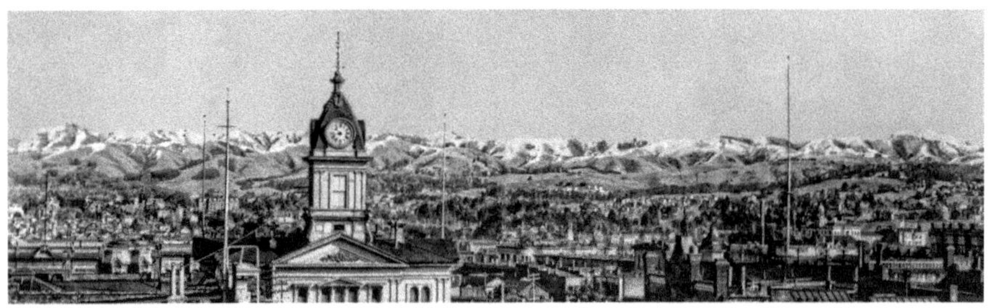

Before nonnative trees were planted in the early twentieth century, the Oakland hills were rolling grasslands. This 1903 photo shows snow on the hills. *Courtesy East Bay Regional Park archives.*

the Anza party in 1775. The oldest surviving map that shows our redwoods was drawn in 1777 by Pedro Font. Equipped with an astrolabe and compass, Font drew the earliest known map of the *bosques* (wetlands) of Alameda, placing the *palos colorados* in their correct location.

Oakland historian Dennis Evanosky has hiked around the hills, carrying a copy of the Anza expedition sketch, in an effort to find the place from which it could have been drawn. He concluded that a hilltop above Leona Heights would have been the best match for the ancient cartographer's west-facing perspective.[33]

From the opposite perspective, looking east from San Francisco Bay, navigator José de Canizares, of the packet ship (*paquebote*) *San Carlos*, made the first map of the San Francisco Bay region in August 1777. The *San Carlos*, with Juan Manuel de Ayala as its captain, entered San Francisco Bay in 1775. Their objective had been to resupply the Anza expedition, but they had been blown off course by storms at sea and were late for the rendezvous.

Each future explorer and seafarer, in turn, sought to improve on the earlier maps and add detail. Captain George Vancouver sailed the British man-of-war *Discovery* to San Francisco Bay in April 1792. He lay at anchor for ten days, the first non-Spanish ship to arrive. Vancouver, along with his local hosts, rode inland on horseback to explore the East Bay and visit Mission San José in Fremont.

In 1816, the Russian brig *Rurik* stopped by San Francisco Bay during the course of a passage around Cape Horn to the Arctic. Its captain was Otto von Kotzebue. Also aboard was the artist Ludwig Choris,[34] as well as naturalists Johann F. von Eschscholtz and Adelbert von Chamisso. In their journals from the voyage, these eastern Europeans expressed a great interest in, and sympathy for, the Natives they encountered during their stay. Choris created many of the earliest and most detailed drawings of Bay Area indigenous people.

Captain Beechey and the HMS Blossom

Captain Frederick W. Beechey first entered the Golden Gate in November 1826 on his sloop HMS *Blossom*, a British man-of-war. Having departed from Spithead, England, in May 1825, the double-hulled *Blossom*, fitted for icy seas, first sailed around Cape Horn and then up the coast of North America and on to the Arctic.

Beechey, one of eighteen children, came from a distinguished family of mariners and artists. His father, Sir William Beechey, was well known as a painter of portraits for British royals. Younger brother Richard Brydges Beechey, who was also aboard the *Blossom*, created many oil paintings of scenes from pastoral California.[35]

Captain Beechey brought transcribed copies of the earlier Spanish maps, as well as information from the voyages of Vancouver and Von Kotzebue.

During the two autumn seasons[36] when the *Blossom* remained in San Francisco Bay, Captain Beechey and his mapmaker, James Wolfe, explored the area. Using a thirty-eight-foot barge, they took hundreds of soundings, from the North Bay to the south.

Upon their departure in late 1827, Beechey compensated their *Californio* hosts, the San Francisco *alcalde* Don Ignacio Martinez and Padre Tomasso of Mission Dolores, by providing them with copies of the new maps and charts.

One odd development resulted from the creation of the Beechey/ Wolfe charts: the names of Alcatraz and Yerba Buena Islands were inadvertently exchanged, leading to some degree of confusion among later visitors. In the Ayala/Canizares maps, the island that today connects the two segments of the Bay Bridge was called *Isla de los Alcartaces*, "Isle of the Pelicans." Today's Alcatraz was named *Yerba Buena*, "Good Herb." After an English transcription error was set in print, the new names became permanent.

The *Blossom* next retraced its voyage around Cape Horn and back to England, where Frederick Beechey remained ashore for the last thirty years of his life. He was actively involved in the Royal Geographic Society of London until his death in 1857.[37]

Blossom Rock

In the San Francisco Bay Area, Captain Beechey may be best known for having discovered and charted its most notorious navigation hazard.

As Beechey discovered it, Blossom Rock was one of the peaks in a ridge of sandstone and clay, as hard as granite, that connects it with sister islands, Alcatraz and Yerba Buena, as well as the smaller Shag Rocks and Arch Rock. At its base, Blossom Rock originally measured approximately 105 feet by 195 feet.[38]

Captain Beechey wrote sailing instructions describing the location of this small island of sandstone, which was about five feet under water. He charted a "hazard line" by which mariners could steer clear of it:

> *As soon as a ship passes the fort* [the Presidio of San Francisco], *she enters a large sheet of water, in which there are several islands, two rocks above water, and one under, exceedingly dangerous to shipping, of which I shall speak hereafter…*
>
> *After passing the fort, a ship may work up for anchorage without apprehension, attending to the lead and the tides. The only hidden danger is a rock with one fathom on it at low water, spring tides, which lies between Alcatraces and Yerba Buena Islands; it has seven fathoms alongside it; the lead therefore gives no warning.*
>
> *The marks when on it are the north end of Yerba Buena Island in one with two trees (nearly the last of the straggling ones) south of Palos Colorados, a wood of pines situated on the top of the hill, over San Antonio, too conspicuous to be overlooked.*[39]

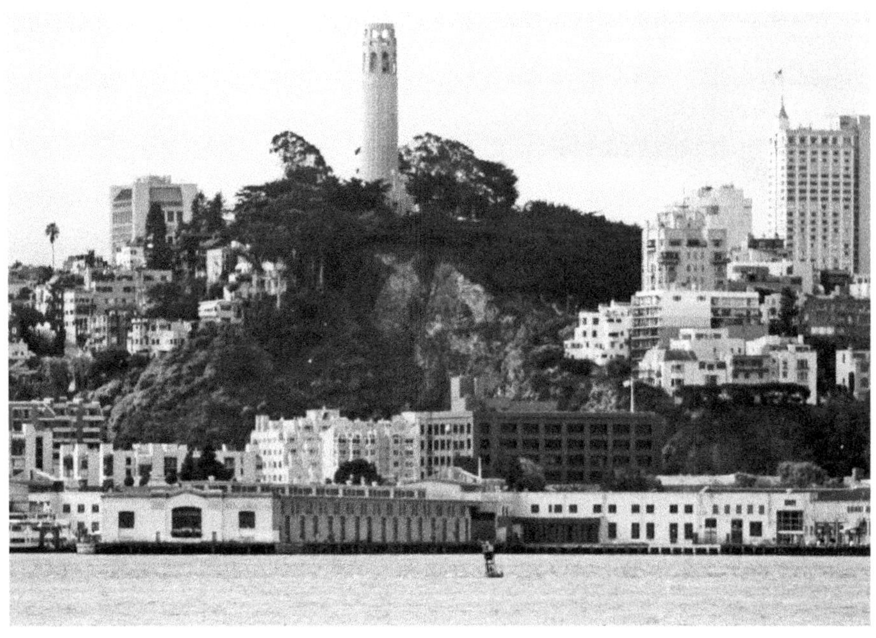

A red buoy bobs above the stub of Blossom Rock. Forester John Nicoles retraced Captain Beechey's hazard line to locate the site of the Navigation Redwoods. *Photograph by Dennis Evanosky.*

Blossom Rock, as the crow flies, is located about one half mile northeast of San Francisco's North Point, near Pier 39. Its position is given as latitude 122 degrees, 24 minutes, 11.16 seconds west, longitude 37 degrees, 49 minutes, 6.96 seconds west in the United States Geologic Survey (USGS) Geographic Information Systems (GIS) database. It was shown on the 7.5-minute "San Francisco North" on USGS topographic maps until 1996.

Since Blossom Rock was named after Captain Beechey's ship, some writers have suggested that the ship ran aground on it. There is no record of this. Beechey probably discovered Blossom Rock as he took soundings from the barge in order to create his nautical chart of the bay.

Captain Ringgold of the USS Colonel Fremont

One American mariner who relied on Captain Beechey's sailing instructions was Cadwalader Ringgold. Having joined the navy in 1819 at age seventeen, Ringgold was soon sent to fight pirates in the West Indies. Appointed captain of the USS *Porpoise*, Ringgold was invited to join the United States Exploring Expedition of 1838–42.

In Fiji, the Americans sought retribution against cannibals who had captured and eaten eleven seamen seven years earlier. After Chief Vendoni of the island of Mololo ordered the killing of two naval officers. Ringgold led eighty men to subdue the natives. Eighty-seven islanders were killed and two villages destroyed, although women and children were spared.[40]

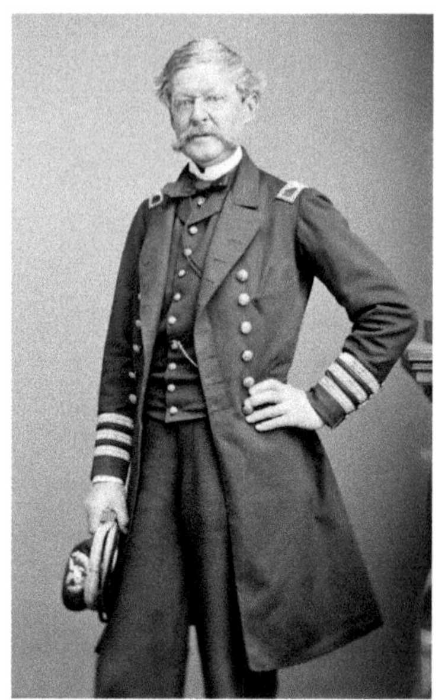

Captain Cadwalader Ringgold sighted and documented the Navigation Redwoods in 1849. After Ringgold sailed to Shanghai, Commodore Perry had him declared insane. Ringgold returned from retirement to active duty during the Civil War and served with distinction. *Courtesy Library of Congress.*

Continuing the expedition, Ringgold took his first voyage to San Francisco Bay. Starting on August 19, 1841, he led sixty men on a twenty-day trek up the Sacramento River as far as Colusa.[41]

Having been promoted to the rank of commander, Ringgold was assigned to further explore San Francisco Bay in August 1849. The American government had taken a great interest in exploiting the recent discovery of California gold.

On this second visit to San Francisco Bay, Ringgold sailed the USS *Colonel Fremont*, a chartered brig. Following the sailing instructions from Captain Beechey, whom he hailed as "the illustrious navigator," Ringgold produced his own chart, showing how to set course to avoid Blossom Rock. He also selected the site that was to become the Mare Island Navy Yard.

Two years later, Captain Ringgold was retired from the navy after contracting malaria in China. In 1854, an official panel, led by Admiral (later Commodore) Matthew C. Perry, declared that Ringgold was "insane" and had him relieved of command.

Ringgold returned to the fleet as a ship's captain during the Civil War. Following his meritorious service, he retired as a rear admiral in 1864 and died in New York three years later.[42]

Destruction

The Navigation Redwoods Are Felled

As of 1850, when Ringgold's sailing chart was published, the tall redwoods used for navigation were still standing in the Oakland hills. By 1855, however, they had been cut down. By 1860, all the ancient redwoods on the west-facing slopes had been clear cut—except for a single crooked tree growing in a steep canyon in Leona Heights.

Blowing the Top Off Blossom Rock, Repeatedly

Despite the availability of navigation charts from Captains Beechey and Ringgold, many ships were still to encounter Blossom Rock.

By 1860, San Francisco had become an increasingly important West Coast commercial port. City authorities sought a plan to demolish the sandstone/clay formation. The chamber of commerce approached the United States Coast Survey to study the best techniques.

In early 1867, Lieutenant W.H. Heuer and Edward Cordell of the Army Corps of Engineers conducted tests with nitrate soda and gunpowder.[43] They concluded that in order to reduce the height of Blossom Rock to twenty-four feet under water, the best strategy would combine hand-drilling with blasting. They recommended using 175-pound charges of powder in watertight casks. The estimated budget would be $56,000.[44] But the U.S. Congress had appropriated only $50,000 for the purpose.

In an effort to get the job done, the City of San Francisco put out a bid[45] for the job, with $75,000 to be paid to the engineer who succeeded in blasting Blossom Rock down to twenty-four feet below the water line. The terms included the proviso "full performance, or no payment."[46]

Civil engineer Alexy Waldemar Von Schmidt[47] submitted a proposal that was accepted. In October 1869, he floated a timber crib with a double-tank cofferdam out to Blossom Rock. The crib was anchored with steel-pointed piles and chains. A shed, accommodating fifteen men for sleeping and cooking, was constructed 50 feet above. Miners, equipped with steel-pointed picks and sledges, dug a tunnel into the rock. By January 1870, it had reached 30 feet below water. Next, with eight miners digging at a time, adding small blasting charges, a cavern was dug. Once the chamber extended 140 feet by 60 feet and was 12 feet high, Von Schmidt publicized the date for his big blast: April 23, 1870. "Immense crowds,"[48] numbering in the thousands, gathered on Telegraph Hill to watch the spectacle. A scow loaded with 21.5 tons of explosives, in thirty-eight sixty-gallon ale barrels and seven boiler tanks, was sunk into the chamber.

Von Schmidt employed another technological innovation: electric detonation, running wires to the explosives, with the water for a ground. Positioning himself on a boat anchored eight hundred feet from the scow, Von Schmidt cranked a magneto battery at 3:30 p.m. A column of water three hundred feet tall shot up into the air, to the delight of the audience ashore. But after the debris had settled, soundings indicated that Von Schmidt had come two feet short of the required twenty-four feet below the water line. The Army Corps of Engineers refused payment until the specification was met.[49]

For the next eight months, Von Schmidt used a floating platform with a chain-operated rake to scrape away rock debris. Finally, he collected his fee. But it is unlikely that he made a profit based on the cost of the post-blast operations.[50]

With the coming of the twentieth century, the drafts of ships had increased. Even in its reduced state, Blossom Rock continued to be a maritime hazard.

In 1903, after three other peaks on the same ridge—Arch Rock and the Shag Rocks—were blasted down to thirty feet below the waterline by the Army Corps of Engineers, Blossom Rock was as well. Finally, in 1932, it was lowered to forty-two feet below the mean waterline.

While Blossom Rock is no longer shown on standard USGS topographic maps, a red buoy remains anchored there to mark the spot where its base remains.

Preserving Knowledge of the East Bay Redwood Forest

William Gibbons

For what we know today about the history of the East Bay redwoods, a large measure of credit is due to William P. Gibbons, MD (1812–1897). He was a physician, naturalist and pioneer of the city of Alameda.

After earning his medical degree in New York City in 1852, Gibbons sailed to San Francisco by way of Panama, arriving in 1853. Very soon after, he became a founding member of the California Academy of Sciences.[51]

Having made his new home in Alameda, in 1855 Gibbons begin exploring the Oakland and Moraga hills, examining the flora and fauna. One of his discoveries was the rainbow trout of Redwood Creek, which he believed to be a new species.[52]

The naturalist was appalled to find the ruins of the redwood forest. Much later, he described what he had seen:

> [J]*ust below and to the westward of the highest point of the Oakland Hills* [is] *a small depression in the hills, about two acres in extent. In…1855, there were about a hundred fifty stumps of redwood, the great majority of which were from twelve to twenty feet in diameter, the trees having been cut from one to eight feet above the surface of the ground.*[53]

Over the next forty years, Gibbons frequently hosted his friends and colleagues on wagon rides to visit to the redwood forest in the Oakland hills. The trees were gradually regenerating.[54]

British naturalist Alfred Russel Wallace (1823–1913), in an 1886 lecture, recounted such a picnic:

> *Dr. Gibbons of Alameda…took me on a drive into the foothills to see the remains of the Redwood Forest that had…been ruthlessly destroyed to supply timber for the city and the towns around. Our companion was Mr. John Muir…*
>
> *We wound about among the hills and valleys, all perfectly dry, till we reached a height of 1500 feet where many clumps of young redwoods were seen, and stopping at one of these, Dr. Gibbons took me inside a circle of young trees from 20 to 30 feet high, and showed me that they all grew on the outer edge of the huge charred trunk of an old tree that had burnt down. This stump was 34 feet in diameter.…The Doctor has searched all over these hills and this was the largest stump he had found, though there were numbers between 20 and 30 feet* [in diameter]*.…These enormous trees, being too large to cut down, were burnt till sufficiently weakened to fall, and this particular tree had been so burnt about 40 years before. We lunched inside this ancient mammoth tree.*[55]

Other scientists who accompanied Gibbons to our redwoods included Professor Joseph LeConte, the UC-Berkeley geologist; Dr. Josiah D. Whitney (for whom Mount Whitney is named), early chief of the California State Geological Survey; and Dr. Alfred Russel Wallace, who shared with Charles Darwin the development of the theory of natural selection.

The journal *Erythea*, in which Gibbons's 1893 monograph was originally published, did not have a wide circulation. Yet his observations of the natural features of the Oakland hills, and the fact that he shared his discoveries with so many colleagues, preserved the information into modern times.

Gibbons and Muir concluded that the redwoods in the Oakland hills had been among the largest on Earth. They believed that the Oakland hills habitat was the place of origin of the coast redwood.[56]

Gibbons knew that sea captains had used the ridgetop redwoods as navigation landmarks:

> *Partly concealed by a dense growth of ambitious saplings on the very pinnacle of the mountain* [probably Redwood Peak] *and overlooking San Francisco and the Golden Gate, are to be found the relics of a redwood tree that once must have been among the most remarkable and gigantic individuals of its species…*
>
> *I have been informed (by our older citizens) that this tree, looming over 300 feet above the mountain pinnacle, constituted a land mark to vessels entering the Golden Gate before lighthouses or other artificial guides to mariners had been established along our coast.*[57]

Origins of the American Conservation Movement

Gibbons became an early conservation advocate. He kept up a correspondence with Muir throughout the 1890s and spoke out in favor of preserving the Oakland redwood forest as a park.[58]

"From the day he first set eyes on the 'ruins of this forest' in 1855, Dr. Gibbons never ceased in his efforts to save the second growth redwoods for park purposes," wrote Frederick J. ("Monte") Monteagle, public information officer for the East Bay Regional Park District, in 1971.[59] Monteagle quoted Gibbons extensively:

> *Even the melancholy pleasure of seeing the ruins of this forest will be lost in a few years unless some protective influence be extended over it.... The fact that is important is the interest which our fellow citizens are beginning to manifest in the construction of a public park which shall be commensurate in magnitude and beauty of the standing of the metropolis of the Pacific Coast.*[60]

A change was underway in public thinking about the California redwoods. At the time of the Gold Rush, they were seen solely as timber, a resource to be extracted. Valuing the natural beauty of the redwood forest from a scientific and spiritual perspective gradually became the predominant viewpoint among Californians.

Building on the work of Gibbons and his colleagues, historian Sherwood Burgess published in 1951 a seminal paper, "The Forgotten Redwoods of the East Bay."[61]

Oakland City Rangers and Naturalists

Twentieth-century Oakland provided ample support for its park department. Dedicated park staff took great pride in the natural history of the San Antonio redwoods in Joaquin Miller Park. Naturalist Paul Covel is best known. Ranger Louis ("Digger") O'Dell is another.

Rich Wirkkala, who served as a Joaquin Miller park ranger from 1965 to 1997, has also studied the large redwoods there. Wirkkala has discovered two large fairy rings, approximately thirty feet in diameter, with remnant stumps in the center. They stand on a high ridge between the Roberts Park archery range and Skyline Boulevard, south of the Chabot Space and Science Center and the Roberts Park residence.

Daughters of the American Revolution Historic Marker

Around the time of the American bicentennial in 1976, the Daughters of the American Revolution (DAR), a women's patriotic organization, erected historical monuments throughout the country. One such marker was placed in the Big Trees area of Joaquin Miller Park. Led by DAR California state regent Drusilla (Mrs. Arthur P.) Strehlow,[62] the monument was dedicated on March 14, 1977. The dedication plaque reads:

> *The Fairy Ring Redwood Trees:*
>
> *Bicentennial plaque marks the historic grove of redwoods in which are found "fairy rings" over thirty feet in diameter. Here grew some of the tallest redwoods in California. These trees were used as landmarks for sailing vessels entering San Francisco Bay in the years prior to compass charting.*

REVERSE-ENGINEERING THE SEA CAPTAINS' HAZARD LINE

Where did the "Navigation Trees," the two tallest redwood trees cited by sea captains, actually stand? John Nicoles may have been the first person to attempt a rigorous analytical solution to the problem of where, exactly, these redwoods must have stood. A registered professional forester, Nicoles served as natural resources manager for the East Bay Regional Park District from 1971 to 1992.

Nicoles's reverse-engineering concept seems simple: Captains Beechey and Ringgold documented a "hazard line" so the navigator could set a course to avoid Blossom Rock. The two known points on the line, until 1850, were "the northernmost point on Yerba Buena Island" and two tall trees at the southern end of the redwood grove on the ridge line, "too conspicuous to be overlooked."[63]

In the absence of the trees, Nicoles reasoned, one could extrapolate the line that connects Blossom Rock to the north point of Yerba Buena Island. Then the point where the imaginary line intersects the Oakland hills would be the Navigation Trees' location.

In the twenty-first century, the location of Blossom Rock is well identified. A red buoy bobs in the channel less than a mile from the San Francisco waterfront, marking the spot between Alcatraz and Yerba Buena Islands. Anyone with a GPS device and access to a small vessel—or an accommodating captain—can cruise there and look eastward toward the ridgeline.

Complication no. 1: Blossom Rock is not a single point—it is a small island, now forty feet underwater. But the USGS has assigned a specific latitude and longitude. For a graphical solution, one can use this point to mark the Beechey/Ringgold hazard line.

Complication no. 2: Where, exactly, is the northernmost point on Yerba Buena Island? The navigation chart from Captain Ringgold's 1850 exploration of San Francisco Bay shows a peninsula of rocks jutting out from the island's north shore. But this feature was dynamited in the mid-1930s when Treasure Island was constructed. Using an estimated point for the second location on the hazard line may be unavoidable.

The distance from Blossom Rock to the Oakland hills is about 12.5 miles. A small change in the angle between two possible sets of points on the bay could create a relatively large position shift on the hill.

Still, even with a less-than-precise line pointing toward Redwood Peak, it is possible to understand what the sea captains were seeing. From a boat in the vicinity of Blossom Rock today, assuming that one is positioned high enough to clear the Bay Bridge, the view through binoculars shows individual trees along the ridgeline. Captain Beechey must have used a spyglass. The chart he created includes a sketch of the ridge top, with the pair of tall redwoods lined up with the north point of Yerba Buena.

The Nicoles Approach

Using a combination of graphical and botanical techniques, John Nicoles is confident that he has solved the mystery of where the legendary pair of redwoods once stood: near the western boundary of Roberts Regional Park, between Skyline Boulevard and the parking lot, near today's Madrone Picnic Area. "I had always assumed that the story of the navigation trees was a myth," Nicoles recalled in 2016.[64]

> *However, one day I was driving up Skyline Boulevard from its intersection with Joaquin Miller Road and noticed that the crown of trees on the ridgetop was substantially taller than those of the surrounding trees.*
>
> *The trees in Roberts Park are not particularly tall, in redwood terms. There are no trees there that are three hundred feet tall. The trees at that location, however, do project above the ridgeline.*
>
> *In timbermen's terms, the Roberts Park site is a "poor" one. A "good" site will grow tall, straight trees. Tree height is a function of "site quality." This is a parameter that incorporates soil nutrients and depth, water retention and solar exposure. The ridgetop sites at Roberts and Joaquin Miller parks produce trees that can be larger in girth but not impressive in height.*
>
> *John Dewitt, a forester who was the head of the Save the Redwoods League,*[65] *was instrumental in shifting the focus from trees of record-setting height to those of great girth. Dewitt doubted Gibbons's report of a stump thirty-three feet wide. He had personally searched the best sites and never found one larger than twenty-five feet.*
>
> *What Dewitt overlooked was that "poor" sites—what we have in the East Bay redwoods in general and near Redwood Peak in particular—tend to produce trees of larger girth relative to their height. The Navigation Redwoods were not necessarily the tallest trees in the East Bay hills, but they were located on the ridge top. And they were tall enough to stick up above their cohorts.*
>
> *When a redwood is felled, shoots grow up from the stump, forming a fairy ring of younger trees. These are genetically identical to the progenitor tree—they* are *the original tree.*

Nicoles observed the ridgeline from various vantage points and saw that there were certain redwoods on the west-facing ridgeline that were clearly the tallest along the ridge: "I zeroed in on the tallest trees that are now visible over the ridgeline. Logically speaking, these would be the descendants of the tallest trees growing there in 1826."

From a distance, Nicoles identified two tall crowns and traced them down to trees that grow between Skyline Boulevard and the Roberts Park parking lot. One is a cluster of three stems sprouted from a single stump. The other is a single stem:

> *These identified trees are one-third again as tall as the surrounding redwoods.…They stick up higher because they have a wind-firm crown, unlike nearby trees.*
>
> *The trees in Joaquin Miller Park may actually be taller and/or bigger in girth, but because they are below the ridge on the hillside, they do not stick up above the skyline. The trees in Joaquin Miller Park generally show less wind damage because they are in a more protected site than those in Roberts that are at the top of the ridge.*

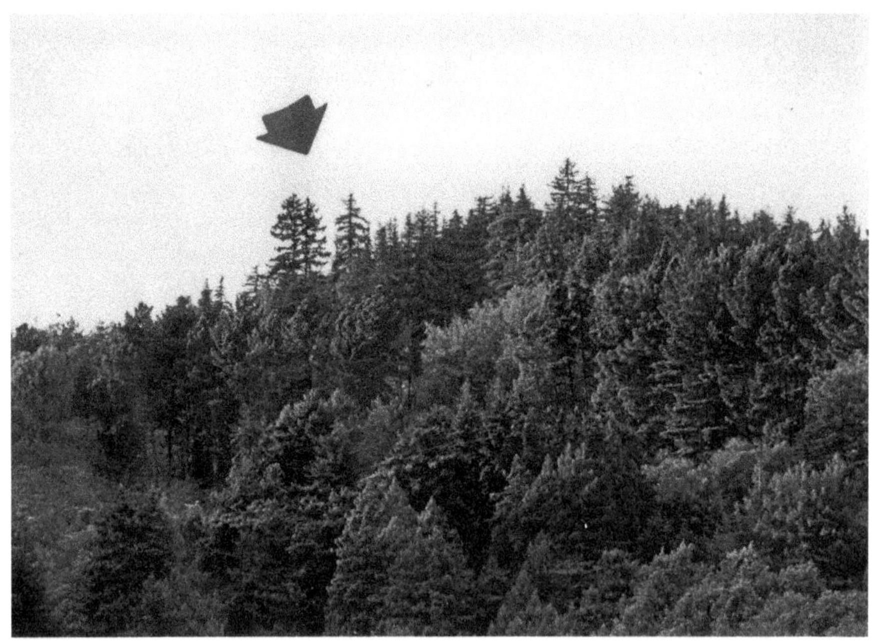

Regional parks naturalists suspect that the two tallest crowns seen above Redwood Peak could be the offspring of the Navigation Redwoods. *Courtesy East Bay Regional Park District archives.*

> *What really convinced me that the trees in this area were the most likely descendants of the Navigation Redwoods was retracing the sea captains' "hazard line" from Blossom Rock.*
>
> *Although it is difficult to do visually now because of the location of the Bay Bridge, when you draw a line from Blossom Rock across the estimated point on Yerba Buena Island and toward the trees at the top of the Oakland hills, you come out exactly at the point where our identified tall trees are growing.*
>
> *The Roberts Park site also agrees with Gibbons's qualitative description of the thirty-three-foot stump location. On any site, "good" or "poor," one would expect the largest trees to have the greatest girth.*

Nicoles can also explain one discrepancy between Gibbons's description of the location of the giant stump and that of today's tallest ridgetop trees:

> *Gibbons described a site that is clearly the Redwood Bowl and indicated that the large stump was located about a half mile southwest of it. The site that I have identified is closer to due south of the Redwood Bowl. I attribute Gibbons's perception to the common misconception that the Oakland hills run on a north–south alignment. They do not. Their alignment is northwest–southeast.*
>
> *Given the clonal quality of the entire stand, such a tree would be readily visible today if it existed in the past. I have to conclude that Gibbons' directional statement was made in error.*[66]

The Minority Opinion

A small but enthusiastic community of local history buffs continues to aspire to pinpoint the exact location of the Navigation Redwoods. While John Nicoles's solution is generally acknowledged as the most rigorous, some people feel that there is room to entertain reasonable doubt. "If you place a straightedge on a map of the Bay Area from Blossom Rock and connect it across the north end of Yerba Buena Island, the line runs right up to where the West Ridge Trail passes the Police Athletic League day camp," noted Rich Wirkkala. "That is exactly where the large stump I observed still exists….If you look at the profile on the map, Redwood Peak is just south of the location that Captain Beechey gave for the Navigation trees. This also corresponds with where the straightedge crosses the West Ridge Trail at the thirty-two-foot stump….Check it out!"

Chapter 4

YANKEES

The habitat of the Coast Redwood, Sequoia sempervirens, *extends…a distance of some five hundred miles…in the Coast Range proper. The only exception to this rule is found just opposite to…the Golden Gate….At this point, the Oakland Hills, being exposed immediately to the sea winds and fogs, bear—or at least once bore—a group of redwoods about five miles square.*
—William Gibbons, The Redwood in the Oakland Hills, *1893*

The Primeval Redwoods, Protected by the *Californios*

Before Europeans came to San Francisco Bay, the extensive redwood forests that grew in wetter centuries past had already receded from their historic ranges. Our East Bay redwoods, along with Muir Woods in Marin County and Big Basin in Santa Cruz, were the only remaining native groves south of the Petaluma River.

The East Bay redwood forest extended only a few miles, from the west-facing hills of today's Joaquin Miller Park and eastward though the ridges and valleys of today's Roberts and Redwood Regional Parks to Moraga Valley. Evidence of the redwood forest that once grew there has been found as far east as the St. Mary's College campus.[67]

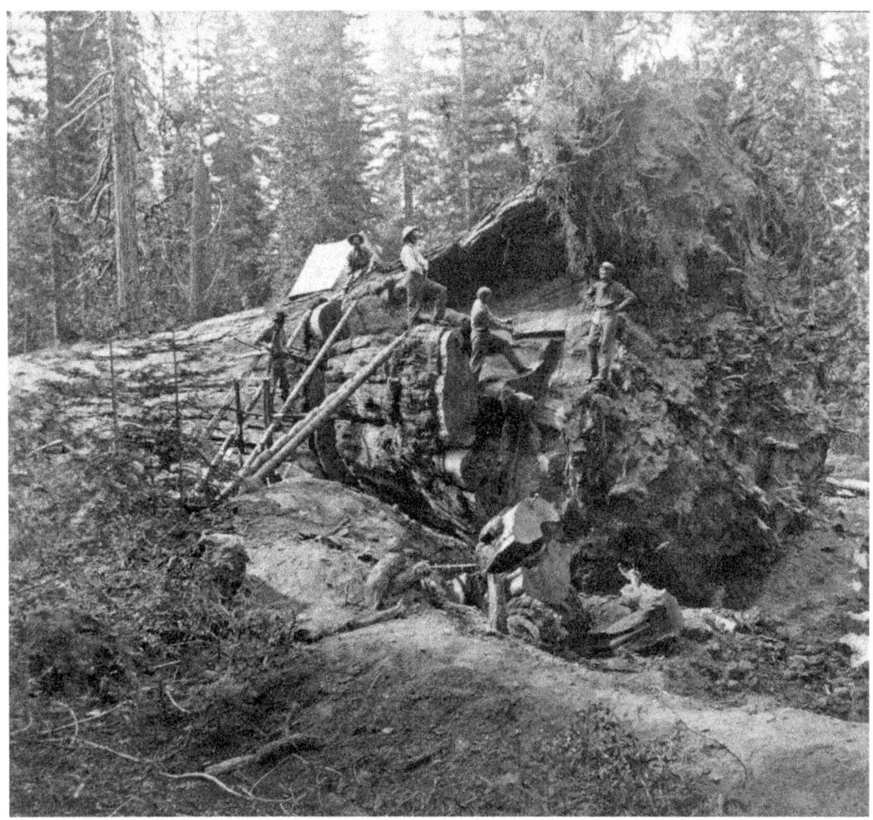

Early naturalists, including William Gibbons, believed that the East Bay hills redwoods could have been the largest that grew anywhere. *Courtesy Society for California Pioneers.*

Names of the East Bay Redwood Groves

In *Californio* times, our *palos colorados* were known in their entirety as the San Antonio Redwoods. The woods were named after the Peralta *rancho*, as well as the waterfront village, over which the Oakland hills appeared in the distance. It was at the foot of today's Thirteenth and Fourteenth Avenues in Oakland where Antonio Peralta had his *embarcadero*.

With the coming of the Yankees after the Gold Rush, the forested districts were renamed. The western slopes—including groves in today's Dimond, Shepherd and Redwood Canyons—retained the name San Antonio Redwoods. But the fragrant valley along today's Stream Trail in Redwood Regional Park became known as the Middle Redwoods after 1849. On the

Moraga side, the watershed lands around San Leandro Creek were known as the Moraga Redwoods.[68]

Until Alameda County was separated out of Contra Costa County in 1853, the boundaries of these forest districts were indefinite. In general, the ridgelines that separated the watershed drainages also marked the divisions between the wooded districts.

The Spanish did not view the redwood forest as a resource to be immediately seized and chopped down. As noted by historian K. Jan Oosthoek, the construction of the Spanish Armada in the 1580s left parts of Spain entirely devoid of trees.[69] Surely this awareness informed the *Californio* perspective.

Besides, prior to experiencing earthquakes, the *Californios* preferred adobe to wood as a building material. The labor required to chop down a giant redwood with hand tools surely must have seemed daunting compared to fashioning fireproof bricks from the clay underfoot.

While the Spanish recognized the value of the redwood timber for construction, there is no evidence that redwood trees were felled for construction of the San Francisco Presidio or Mission Dolores. The East Bay redwoods did provide the lumber for Mission San José, though.[70] When Luís Maria Peralta received his land grant in 1820, the Catholic Church retained what would now be called timber rights to the *palos colorados*.[71] The text of the land grant, as written by Father Narcisio Duran, pastor of Mission San José, prohibited felling the redwoods for commercial purposes. The 1835 land grant to Joaquín Moraga and Juan Bernal contained similar language.

Under the terms of the grants, any lumber harvested would "at all time be at the disposal of the Missions," and any cutting of wood must "remain in common." While the terms of the land grants were written by the church, it would appear that the Spanish monarchy had originally intended to preserve the redwood forest as a resource for future ship building.

Emigrants Come to the San Antonio Redwoods

In the coastal towns of California, traders and itinerant Europeans had landed by the late 1830s. Among the first non-Spanish Europeans to arrive in East Bay redwoods were sailors jumping ship. Over the objections of Antonio Peralta and his neighbors, illicit logging was underway in the redwood forest by 1840.

George Patterson, an Irishman, deserted the English barque *Columbia* in San Francisco harbor and went to San Antonio. John Parker deserted the English ship *Sulphur* that year. Patterson and Parker sold the lumber they hewed to John Sutter.

Mofras noted in his diary, while visiting San Francisco in 1841, that French naval deserters Sicard and Leroy were logging the redwoods.[72] It is likely that one of them began construction on the Palo Seco Mill in 1841.

For the next few years, logging was sporadic, influenced by fluctuations in demand for lumber and the price it could command. There are no records of logging in the redwoods between 1842 and 1846. After Mendenhall and the Smith brothers had been driven out by the cavalry, their employer, John Sutter, invested in Sierra sawmills instead. Thus, demand for East Bay lumber temporarily ceased.

Occasional logging was reported in the San Antonio redwoods in 1846 and 1847. There was a hiatus in 1848—the denizens of the redwood forest had headed for the gold country.[73] Logging in the San Antonio redwoods began in earnest after unsuccessful gold seekers returned in 1849.

Residents of the East Bay Redwoods in the 1850s

What we know about the men—and a few women—who were tough enough for life in the redwoods comes from a sparse collection of stories. Newspaper articles tell of the depredations of "the notorious mob element from the Redwoods" when they were riled up by livestock rustlers enough to demand justice on their own terms.

Yet the "Redwoodites" formed a community based on mutual assistance. The mill operators, doing business among themselves and with lumber buyers and storekeepers, needed to be mindful of fair dealing. A dishonest person would readily acquire a bad reputation in the small, insular and rough society in the woods.

"Monte" Monteagle: The Newsman Who Popularized Redwood Canyon History

Journalist Frederick J. "Monte" Monteagle may have done more than anyone to keep the lore of the 1850s East Bay redwoods in circulation into the present century.

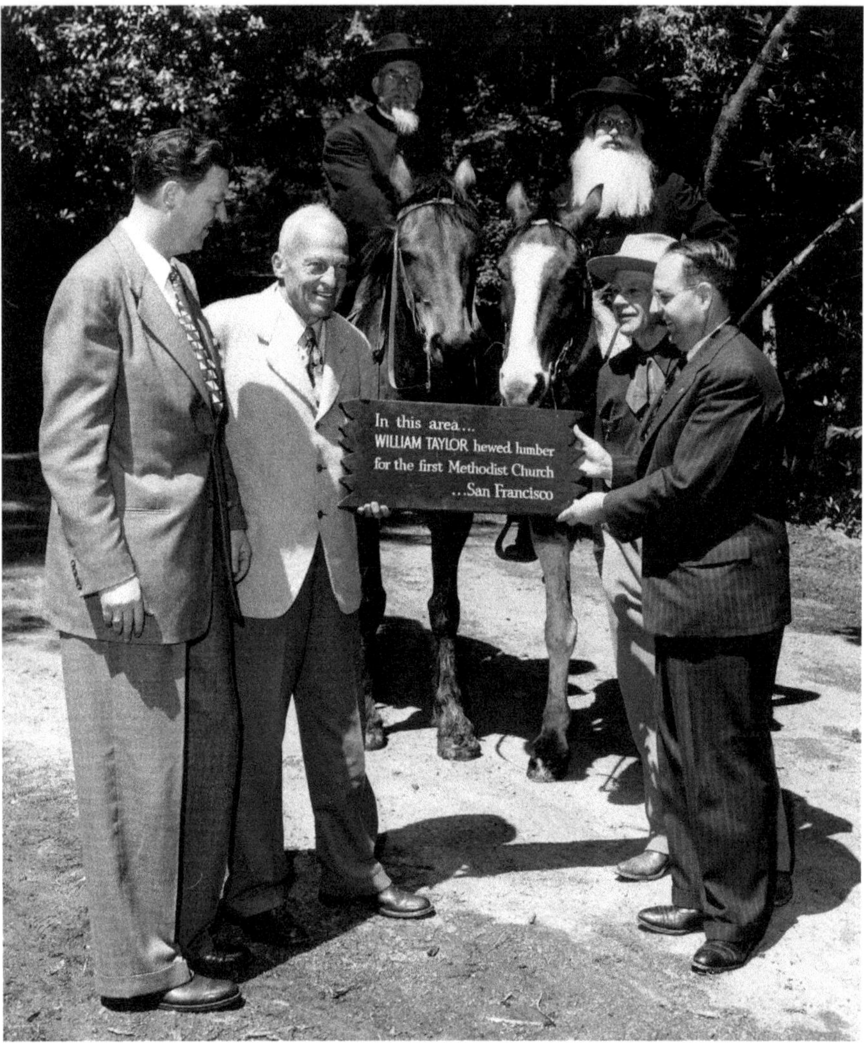

Regional Parks officials celebrate the Redwood logging centennial in 1953. *From left to right*: EBRPD general manager Richard Walpole; Robert Sibley, EBRPD board president; two mounted actors; Oakland naturalist Paul Covel; and an unidentified man. *Courtesy East Bay Regional Park District archives.*

Capping a long career spent writing for the *Oakland Tribune* and other newspapers, Monteagle became the park district public information officer in about 1970.

Monteagle familiarized himself with the history of the "Redwood Boys." He studied nineteenth-century clippings from the *Tribune* archives,

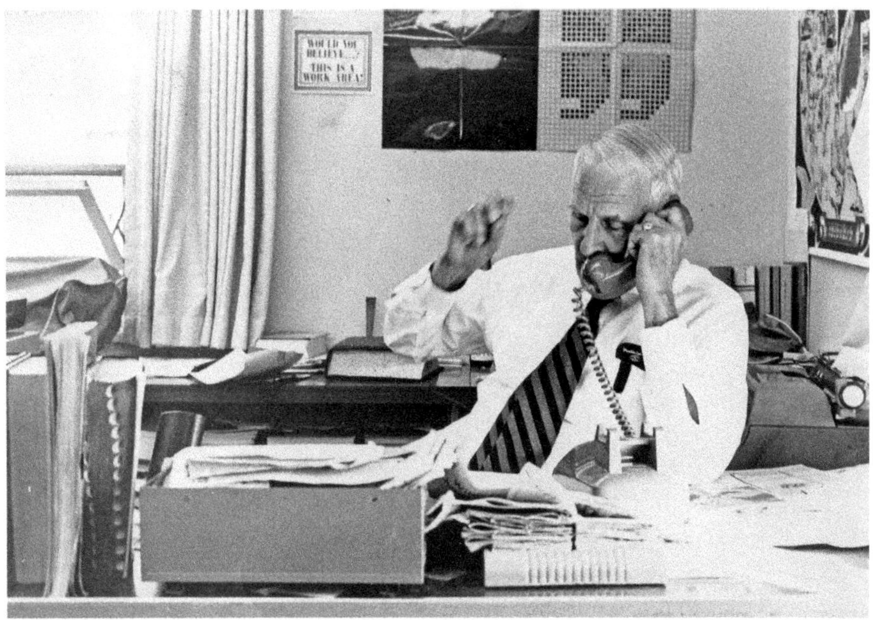

Journalist Frederick J. Monteagle studied logging-era news reports to compile *A Yankee Trader in the California Redwoods* in 1976, publicizing the history of Redwood Regional Park. *Courtesy East Bay Regional Park District archives.*

as well as memoirs by Joseph Lamson and other documents. Compiling the wildest tales from his various references, Monteagle retold them in a series of feature articles. The park district collected them into a booklet, *A Yankee Trader in the California Redwoods*, which was published in 1976.

Ned MacKay, who joined the park district's Public Information Office in 1982, knew Monteagle. "Monte prided himself on his florid writing style," MacKay recalled in 2016. "But he adhered to the journalist's ethic for accurately reporting the story."

Joseph Lamson

The most detailed portraits from this unique time emerge from the diaries of Joseph E. Lamson.[74] The widower left his beloved daughter, Anne, at home in Lubec, Maine, and set off for the California gold fields to seek his fortune. As he journeyed across the sea and throughout California, Lamson described his travels for Anne in sporadic journal entries.

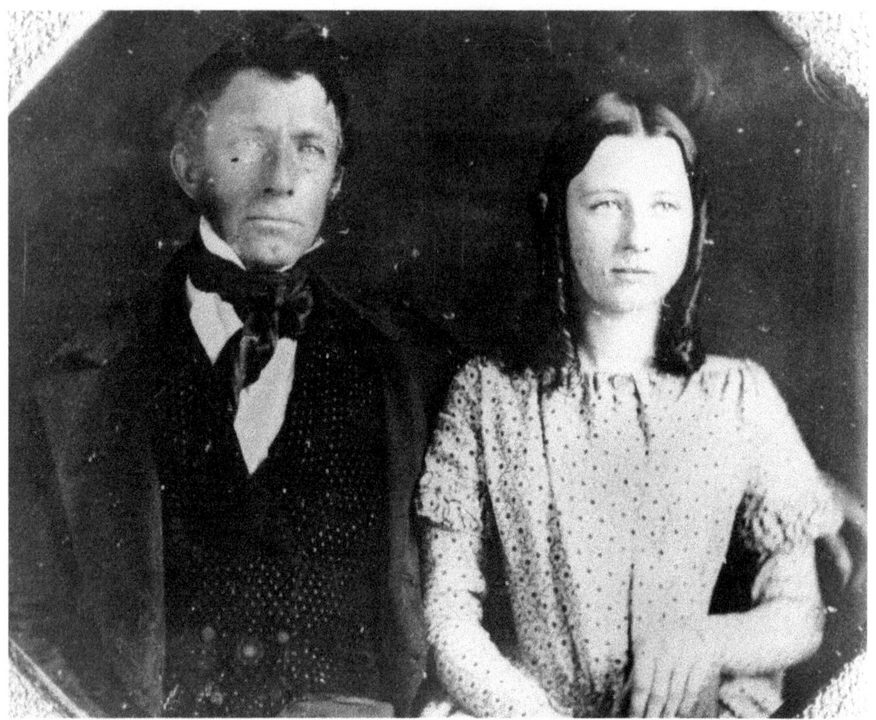

Joseph E. Lamson recorded the story of 1850s Redwood Canyon in a diary for his daughter, Anne, who stayed home in Maine. *Courtesy East Bay Regional Park District archives.*

Lamson landed in San Francisco in 1852. The following summer, he made his way to the East Bay redwoods. Within the year, he was operating a combination lodging house, general store and saloon catering to the loggers. Lamson was an amateur naturalist who wrote articles for *Hutchings California Magazine*.[75] He kept rattlesnakes and vultures as pets.

Lamson wrote seven journal entries from what he called "the red woods, Contra Costa." The first such entry, dated December 1853, describes his feelings of loneliness after all the trees in his immediate vicinity were cut down and the loggers decamped:

> *Three or four months ago I was surrounded by a deep, dense forest, in which was a busy population at work. But this industry fast swept away the forest, and as the timber grew scarce, they began to remove to other places. They continued to go until our society was reduced to ten men, living in a little cluster of four cabins. But even this colony has taken a sudden*

resolution to migrate, and this morning the last man went, and I am left alone. So now, nothing remains for me but to go too, which I shall do as soon as I can determine where.

The Bullwhacker Way of Life

Prior to the California Gold Rush, newcomers to the redwoods were often sailors who had abandoned ship. The rigors of hacking down giant redwoods or gently persuading ox teams must have seemed preferable to the hazardous life at sea. "I have seldom witnessed such barbarous cruelty or such an absence of skill in driving," Lamson wrote.

To a man accustomed to the mode of teaming in Maine the manner of driving oxen in California is matter of much curiosity. A great portion of the teamsters that I have seen are from the western states. These men use a whip made of strips of green hide braided into an immense thong, attached to a rough handle with the barks on. These whips, lash and staff, vary in

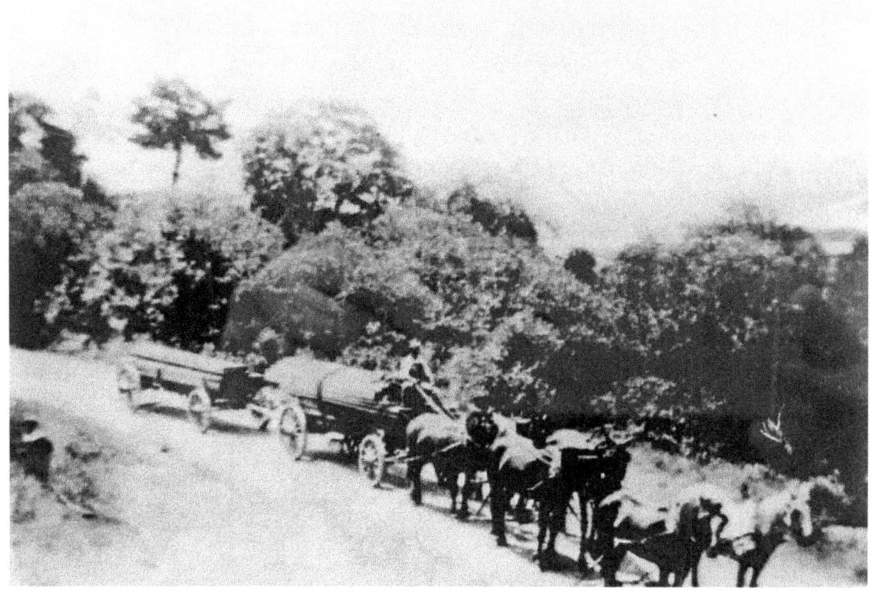

A team of horses hauls rough-milled lumber from the Middle or Moraga Redwoods south toward Castro Valley. *Courtesy Doris Marciel.*

> *length from 12 to 20 feet. The drivers have the art of cracking them with a report, I have never heard equaled by a stage driver and when half a dozen of them are driving their teams over a bad road, the report of their whips resounds through the canyon like an irregular discharge of musketry. They are cruel instruments of torture and are used without mercy.*[76]

"According to the accounts of timid observers, the redwoods were in many ways a suburb of hell populated by ship deserters…who remained undefiled by temperance or cotillion, and were tough and unregenerate to the end," wrote Monteagle. "[T]he feared, hard-drinking, oath-bawling loggers, dubbed the 'redwood boys,' 'redwoodites,' or 'redwood rangers' by an awed Bay Area press, would emerge from the forest fastness from time to time to deal out summary and retributive justice to horse and cattle thieves."[77]

Lynching Rustlers

According to Monteagle, at least four lynchings were perpetrated by the "redwood boys" in 1854 and 1855. The mob violence was provoked by thefts of the beasts on which the men depended for their livelihoods.

Lamson, in his diary entry of August 17, 1854, described the circumstances of his two neighbors losing four oxen and a horse between them:

> *The great facilities for concealing oxen, horses, and other property in the innumerable deeply secluded valleys and hiding-places that occur in every direction in the mountainous country…commencing at these Redwoods… offer too many inducements to the numerous idlers and vagabonds that prowl about…and consequently theft, robbery, and I may almost add, murder, are but every day occurrences. No man who owns a horse, an ox or swine, can feel secure of them for a moment when out of sight. These thieves are often associated in large gangs, and consist of both Americans and Mexicans; Many of the butchers are supposed to be league with the thieves.*

Lamson accompanied his neighbors to San Antonio for vigilante action against the rustlers. His account of the lynchings of Amande Careene and Peter Auchimbault is corroborated by an article in the *San Francisco Daily Evening News* from August 23, 1854.

Livestock rustling had been widespread in the village, the paper reported. The crime wave had been vexatious for several weeks. The constable of

DAILY EVENING NEWS.

OFFICE—NO. 85 LONG WHARF, CORNER OF FRONT ST.

W. & C. JULIAN BARTLETT, EDITORS.

Wednesday Evening, August 23, 1854.

Publication Office.—The Publication Office of the EVENING NEWS AND PICAYUNE, is removed to No. 35 Commercial street, corner of Front, where all Advertisements and orders for Printing should be left.

Agents for the Daily Evening News.
M. ISAAC ROSENBAUM, Stockton.
Messrs. STILES & DODD, (Periodical agents) Benicia.

The Daily Circulation of the Evening News is nearly Double that of any Evening Paper in the State.

"Justice to Frenchmen."

Under this head, we notice several days ago in *Le Messager* a communication sent from Mariposa and said by the editor, in his accompanying comments, to have been signed by sixty-seven Frenchmen. This communication struck us at the time as unjust, and calculated to have mischievous effects—and as we see it to-day translated into English and copied into one of the morning papers, we will endeavor to show in what its injustice and mischeivous tendency, according to our opinion, consists.

This communication complains, and evidently with justice, of several brutal outrages that have been perpetrated upon Frenchmen at the Mariposa Camp; and also of the failure of the officers of justice to sufficiently punish those who committed them. So far we grant the writers are correct. If the outrages they recount were committed—and if the authorities permitted the perpetrators to escape with the light punishment stated, which we fully believe, not only because these gentlemen so state, but because of the exceeding probability of the facts—the Frenchmen of Mariposa have a right to complain of the injustice through the press. Such outrages should be made public; and when justice is refused at the hands of the judge, an appeal should be taken to the high court of public opinion. But what we think wrong, is the attempt made to create the belief that it was owing to the fact that these parties were *French* that this injustice was done them. When our papers from the interior are constantly filled with the accounts of lawless acts—when nothing is more common than to hear of outrages similar to those of which the correspondents of *Le Messager* complain—and when the general impunity of crime in California is a by-word—we cannot see how the French portion of the population can reasonably expect to escape, any more than the American portion, from the acts of outlaws—and we do think it unjust in them to attempt to raise a belief that a distinction is drawn against them *because* of their nationality; and that they are robbed, or beaten, or murdered, merely from the fact that they are Frenchmen.—Such is not in our opinion the case. If such outrages were *confined* to them, we would admit that there was some justice in such a belief. But, as every body knows, such is not the fact; Ameri-

Lynch Law in San Antonio!!
TWO MEN HUNG!!!

The usually quiet little village of San Antonio on the opposite side of the Bay, was early this morning the scene of another sad and fearful instance of retributive justice under the summary code of Judge Lynch.

The circumstances of the case, as near as we can ascertain them, are as follows:—For several weeks past, the residents of the San Antonio *encinal* and vicinity, has been troubled and harrassed continually by cattle thieves, who would visit their premises at night, and drive off all the most valuable stock they could find; fine milch cows, oxen, calves, etc. have thus disappeared from the corralls, and for the time being, all trace of them or the "lifters" lost.

Yesterday, however, the constable of Clinton Township, Mr. Carpenter, received information which led to the development of the whole matter. The culprits proved to be two men, one, a Frenchman, named Amande Careene and the other, a native of Illinois, but recently from Cincinnati, Ohio, known as Peter Auchimbault

They keep a slaughter yard near San Antonio, in which were found several head of the stolen animals, three alive, and some four or five others killed. Upon examining the yard carefully, the skins of two cows, stolen last week from a person living in the town, were discovered in a deep hole, which had evidently been dug for the purposes of concealment. In the evening the guilty parties were taken in custody, and placed under charge of deputy constables in Wetherell's Hotel to await examination before a Justice of the Peace.

Great excitement ensued upon the citizens learning the foregoing facts, and a short time after the arrest, a meeting was held by the Vigilence Committee, recently formed in Oakland, and others, to deliberate as to the expediency of immediate punishment by death. It was decided by a majority of one, that the law should take its course, and the accused be allowed to remain with the legal authorities.

During the night, however, a pary of men, some forty or fifty in number, arrived from the red woods,—another meeting was held, and at six o'clock, this morning, the prisoners were taken to a place in the rear of the Mansion House to an oak tree and there hung.—The Frenchman is reputed to have been quite wealthy, and a large sum of money was offered by his country-men who live in San Antonia, to save his life. The offer proved useless as did also a forcible attempt to rescue him. Archambault, we understand, has left a wife and family to mourn his loss and sorrow for his disgrace.

Whig Nominations.

The following are the nominations made last night, by the Whig City and County Convention:
For Mayor—J. P. Haven.
City Tax Collector—Edward T. Batturs.
City Treasurer—Wm. Neely Johnson.
Recorder—W. R. Turner.
City Attorney—Lorenzo Sawyer.
Street Commissioner—Edward Ebbets.

The nomination of Mr. Haven for Mayor, was made on Monday evening, but owing to rumors of unfairness, in the balloting, Mr. H. sent in a note declining to accept the nomination, unless the gentlemen who competed for it, were satisfied and convinced that no taint rested upon the action of the Convention. The communication, after considerable discussion, was accepted as a resignation, and the Convention proceeded to make a second choice. On the 1st ballot, Mr. Haven was again

The Agricultural Fair, and Premium [...]bies.

We lay before our readers the following [...]ple of a description of letters that editors a[...] ten honored with. They will perceive that [...] simple announcement of what we thought w[...] be an interesting bit of news, viz: that our [...]ricultural Society had offered prizes for the f[...] Babies, we have lost the favor of our patron, Jones; and as an offset, have made a fast fr[...] of his wife, Mrs. Jones. We are very sure w[...] not losers by this transaction—for with su[...] friend at court, as the lady, we need not fea[...] that the man Jones will come over, and again [...]scribe; we should lose our confidence in the [...] else. We, as by gallantry bound, of course, t[...] that the lady is perfectly right in contesting fo[...] prize, and hope she may carry off the *first* h[...] As to the place of meeting and the other par[...]lars, we refer our correspondent to the edit[...] the *Cal. Farmer*.

MR. NEWS:—In a late number of your paper, a list of premiums offered by the "California Agricultural Society," among which I noticed tw[...] the two best specimens of babies. Now, Mr. *Ne*[...] am a woman, and of course, take a great interest i[...] "little innocents" and I am right glad to see thi[...] dence of interest shown in them by the "lords of [...]tion."

You must know, dear sir, that I intend to hav[...] little one on hand at the time appointed, and I a[...]tain, that if the judges decide fairly, that I'll tak[...] of the prizes. You may think this partiality in me[...] every one who has seen my boy, declare that he is [...]markably handsome child. It was only yesterday Mr. B. (who is a candidate for an office, and cal[...] our house to talk to my husband about politics,) de[...]ed that never had he seen such a fine boy. So l[...] for his age, (he will be ten months in Sept.,) with fine features and such beautiful eyes! and withal a bright, intelligent look!

Now, Mr. *News*, (although I'm certain of wi[...] the prize,) Mr. Jones objects to "putting his son on[...]bition, as if he was a prize pig or a fatted calf, poked in the ribs by sour old bachelors, and pinche[...] interested grand-mothers," and declares that he wi[...] listen to such a thing for a moment. But I am d[...]mined to enter my little one, and take a premiu[...] "Competition," says Adam Smith, "is the life of tra[...] "and hence we find that where there is competition i[...] manufacture or culture of any article, it is there ca[...] to its greatest perfection." The offering of prem[...] for the finest stock, etc, in all countries, has prod[...] the most beneficial results. Why should not this [...] to be the case with regard to babies? I can se[...] good reason, Mr. News, and I have tried to argu[...] point with my husband, but its no use, for every ti[...] stump him in an argument, he hotly declares "[...] women know nothing about such things," and th[...] abominates "female Websters." What do you t[...] about it? Don't you think it is natural and right me to *insist*, on having our "little responsibility,' the exhibition? If so, please let me know the [...] where the exhibition is to be holden, and if you sh[...] attend, look out for a bright, flaxen-haired, blue-[...] youngster, you will find him in the arms of his mo[...]

BETTY JONE[...]

SAN FRANCISCO, Aug. 23d, 185[...]

The boys from the redwoods made big news when they pursued livestock rustlers into Oakland and formed a lynch mob. *Courtesy East Bay Regional Park District archives.*

Clinton Township, Mr. Carpenter, had received a tip that Careene and Auchimbault were keeping stolen animals in their slaughter yard in San Antonio near the waterfront. Constables duly arrested the pair. They were held under guard at Wetherell's Hotel to await the arrival of the justice of the peace:

> *Great excitement ensued upon the citizens learning the foregoing facts, and a short time after the arrest a meeting was held by the Vigilance Committee, recently formed in Oakland, and others, to deliberate as to the expediency of immediate punishment by death. It was decided by a majority of one, that the law should take its course, and the accused by allowed to remain with the legal authorities.*
>
> *During the night, however, a group of men some forty or fifty in number, arrived from the red woods...at six o'clock this morning the prisoners were taken to a place in the rear of the Mansion House to an oak tree and there hung.*[78]

Historian Dennis Evanosky noted that the location of the Careene and Auchimbault abattoir was approximately the same as today's Burger King on East Twelfth Street, just south of Fourteenth Avenue in Oakland.

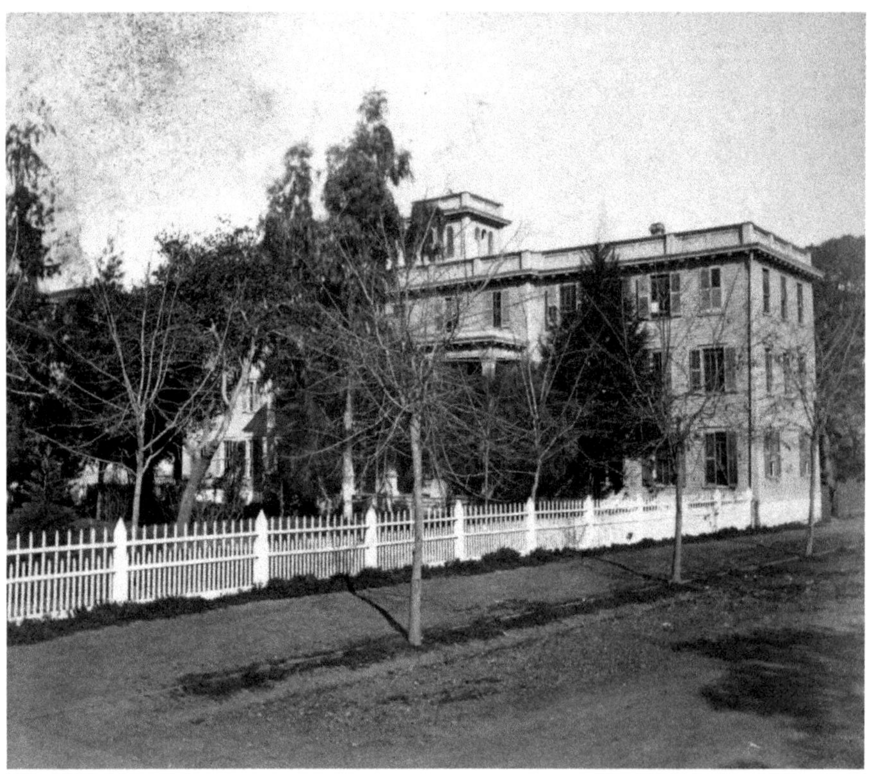

The Mansion House Hotel, located at the northeast corner of Fifteenth Avenue and East Twelfth Street, was the first location of the College of California. The hangman's tree used by the Redwood Boys was behind the hotel. *Courtesy Society California Pioneers.*

As for the lynching, civic discussion followed for several weeks. "Many of the people of Oakland are highly exasperated at the audacity of the Redwoods boys, and threatened to go hang them from their own trees. But this served rather to amuse the boys than to frighten them," Lamson wrote.

Soon thereafter, three oxen were stolen from mill owner and road builder Hiram Thorn. A group of "redwood rangers," an estimated 250 strong, marched up and down the streets of Oakland carrying rifles:

> *A few weeks after these executions, word was brought to the Redwoods that a poor man had been robbed of some oxen in Oakland through the villainy of one of the officials in that city. A company quickly assembled and marched down to the city, determined to have justice done the poor man, and hang the officer if circumstances required it.*
>
> *They had not forgotten the threats of the Oaklanders to hang them, and determined to put their courage to the test. The case was investigated by the mayor of the city, and the mob resolved to await his decision. But much time was occupied in the investigation, and they grew impatient and clamorous.*
>
> *Meanwhile many of them paraded through the streets, uttering defiance to the citizens. "Here is a target," said a brawny, black-bearded Kentuckian (the same I had encountered in the Redwoods, and who sold me a vulture), as he strode along with a rusty rifle on his shoulder, and struck his breast. "Here is a target for the Oakland sharp-shooters. Let 'em try it if they dare."*
>
> *"I'm from the Redwoods," roared out another. "Where is your Oakland company to hang me?" "What are you after?" asked a spectator of one of the boys. "Justice," he replied. "But how are you going to obtain it?" "By the halter, if the money isn't paid pretty soon," he replied with an oath.*
>
> *The affair was approaching a crisis. The mayor's investigation had been protracted, and the clamors and shouts of the mob often reached his ears, when at last he found it necessary to acknowledge that the proceedings of the officer were illegal, that the city was liable for the value of the cattle, and in order to appease the mob, he pledged his individual word for the payment of the money. The party then returned triumphantly to their homes in the Redwoods, and thus the affair ended.*

Thus, Mayor Horace Carpentier promised that the City of Oakland would make Hiram Thorn whole again.

Religion in the Redwoods

Religion first came to the East Bay redwoods in the person of Methodist minister William Taylor. Taylor, who was known for having walked across the continent of Africa, brought his missionary zeal to California as well. Arriving in San Francisco in 1849, he soon ascended a carpenter's bench in Portsmouth Square and preached the "glad tidings of salvation." Men rushed out of the gambling houses to see the spectacle. He later estimated that he had preached one hundred sermons while standing on whiskey barrels.[79]

Discovering that his church stipend would not cover the cost of San Francisco housing for him and his wife, as well as their daughter, Ann, Taylor took a boat to the East Bay and hiked up the hill to cut three thousand redwood shingles. These he traded for enough lumber to build his house at Jackson and Powell Streets. Nearby, he built a church. Then, on the waterfront, he built a seamen's bethel (a chapel).

While in the redwoods, Taylor preached the first sermon to be heard in these parts. He recalled the day in his diary entry of Sunday, October 21, 1849:[80]

> *For retirement and meditation I have strolled out to the top of a high hill.... Looking eastward I see a dense forest of huge redwood timber...west and north, hills and mountains stretch to the uttermost line of the ken of vision, and the scene in its barrenness and sterility, is only relieved here and there by a small oasis, or by the herds of cattle feeding on the dry grass. Southward the whole valley, for fifty miles, is filled with fog...settled over the whole region of the bay of San Francisco....A little to my right are two graves. There sleep the dust and buried hopes of two California adventurers.*[81]

Thereafter, Reverend Taylor preached to a congregation of about twenty-five loggers, about a quarter of the population of the redwoods in 1849.

The A.R. Andersen tombstone, and that of an unknown companion, were hewn from sandstone boulders. The site is within the archery range in today's Roberts Regional Park. In light of the persistent issue of historic sites being vandalized, regional parks staff are instructed to not disclose exact locations of these graves.

A second preacher was described by Joseph Lamson. William J. Brown was a Missouri Campbellite minister from Pike County. "Parson" Brown rode a mule. He split rails, posts, pickets and shingles during the week and

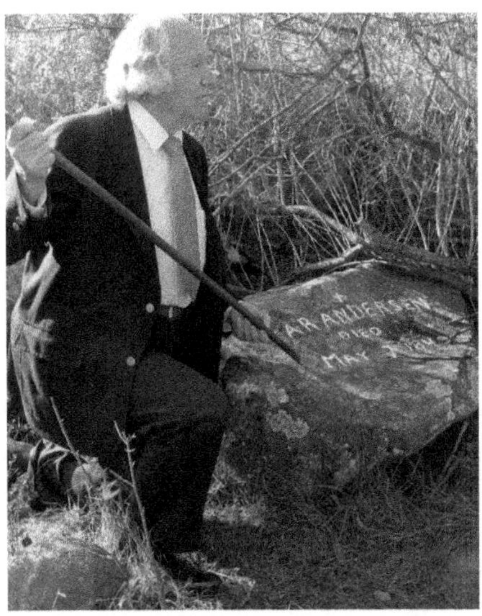

Methodist bishop Donald H. Tippett in the early 1950s visited the pioneer grave site discovered by Methodist bishop William Taylor in 1849 on the day he preached the first sermon to the redwood loggers. *Courtesy East Bay Regional Park District archives.*

preached under the redwoods on the Sabbath.[82]

Monteagle recalled the legend of how Parson Brown handled a man who tried to cadge a drink on the boat bound for Oakland. Brown "sent him sprawling to the deck and would have stomped him if women had not shrieked.…This little affair very naturally increased the parson's popularity in the redwoods."[83]

Lamson described the observances of Christmas 1854 held by the men from Pike County: "There was a large company assembled, and they had plenty of whiskey and a glorious time. The host got beautifully drunk and became very pious; and to wind up his Christmas festivities in the best possible style, he came down to Parson Brown's, the fighting Campbellite Pike County minister; had a meeting called, made a public profession of religion, and was baptized and taken into the church."[84]

"Kanaka Joe" and Hannah Tracy

Of all the neighbors mentioned in the Lamson diaries, none is a more compelling figure than Hannah Tracy.

Following the customs of women in her native Hawaii, eighteen-year-old Hannah rode among the redwoods astride her horse, clad in a long blue calico dress. Nineteenth-century British adventurer Isabella Bird Bishop described Hawaiian women on horseback in 1872:

> *Saturday afternoon is a gala-day here and the road was so thronged with brilliant equestrians, that I thought we should be ridden over by the reckless rout.…The women seemed perfectly at home in their gay, brass-bossed, high peaked saddles, flying along astride, bare-footed, with their orange and scarlet riding dresses streaming on each side beyond their horses' tails,*[85] *a*

bright kaleidoscopic flash of bright eyes, white teeth, shining hair, garlands of flowers.[86]

Lamson praised the equestrian skills of his neighbor Hannah, as well as her personality and good character: "She is kind and affectionate, very lively and excitable, quick and passionate, simple and guileless…she is honest and trustworthy, faithful to discharge all debts she may contract, and fulfills all her engagements."[87]

On her visits to the store on business, Hannah had lengthy conversations with Lamson. Intrigued by the first Sandwich Island woman he had ever encountered, the storekeeper answered her many questions about his family back home in Maine. Lamson attempted to capture in his writing the pidgin dialect of her speech. He showed her a daguerreotype image of his daughter, Anne. Hannah smoked cigars, Lamson wrote, but took no strong drink.

At age eighteen in 1854, Hannah had for five years been married to Joseph Tracy, a sailor from Maine. Lamson struck up an acquaintanceship with his fellow expatriate. "Kanaka Joe" told Lamson of his service on the steam warship USS *Princeton*. On February 28, 1844, the newly built ship had been put on display by the docks at Alexandria, Virginia. Joe was on board.

The captain decided to fire the ship's longest gun, "the Peacemaker," to impress the dignitaries. When the gun exploded, six people were killed, including members of President John Tyler's cabinet. Tyler himself was spared; he was below decks, drinking toasts with American-made whiskey.

Thereafter, Joe sailed the Pacific Ocean. He took a great liking to Hawaii and presumably married Hannah there. The couple planned to return after Joe had made his fortune in California.

Lamson Departs from the East Bay Redwoods

The last entry Lamson wrote from the East Bay redwoods, wherein he describes attending the *fandango* at the home of Joaquín Moraga, was dated January 1855. The next entry places Lamson near Mount Shasta.

After returning to Maine, Lamson published his account of his travels in 1861.[88]

He also pursued an artistic career, drawing and painting the landscapes he recalled from California.[89] He died in Sebec, Maine, on January 1, 1894.

Short-Lived Voting Precincts in the East Bay Redwoods

Alameda County was partitioned out of Contra Costa County in 1853, with the county line drawn to evenly divide the redwood timber resource between the two counties. At that time, the population in the redwood forest outnumbered that of the villages of San Antonio, Brooklyn and Oakland. Logging was the richest source of economic activity in the two rural counties.

Election records from 1853 show three to four hundred men in the redwoods prior to the creation of Alameda County.

Until all the redwood trees were cut down, each of the two counties had a voting precinct in the redwoods. The Prince Mill was the 1854 polling place for Contra Costa County. Tupper's Mill served that function for Alameda County. In later years, it was the Brown and Eager Mill.

Burgess estimated that 50 loggers worked for each of the four largest mills. This is roughly consistent with the estimated 250 men from the redwoods who descended on Oakland in the summer 1854 in pursuit of livestock rustlers.

After a few years, the loggers were gone. The Contra Costa County election precinct in the redwoods was shut down in 1857. Its Alameda County counterpart was no more by 1860.

Chapter 5

LOGGERS

These fellows are charming proficient in the art of swearing, and the beauty of their oaths and the skill with which they use them would make a Penobscot lumberman blush at his ignorance.... The names they give their oxen have rather an amusing sound to an Easterner's ear.... They frequently honor them with the Christian names of their friends.

—*Joseph Lamson,* Round Cape Horn: Voyage of the Passenger-Ship James W. Paige from Maine to California in the Year 1852

William Mendenhall, the Smith brothers and Elam Brown may have been the first American immigrants to log the East Bay redwoods on a commercial scale. Using hand tools and ox carts, they hauled the lumber down Dimond (Sausal Creek) Canyon to the San Antonio wharf. "It was during this period that Elam Brown met William A. Leidesdorff, who sold him his Rancho Accalanes (Lafayette)," the *Oakland Tribune* reported, quoting Burgess.[90]

Sawmills in the San Antonio Redwoods

Construction began in 1841 on the two first sawmills in the San Antonio redwoods. Nathan Spier and Captain William S. Hinckley constructed a water-powered sawmill that turned out shingles and planks, beginning in 1841.[91] The location of this mill remains unknown.

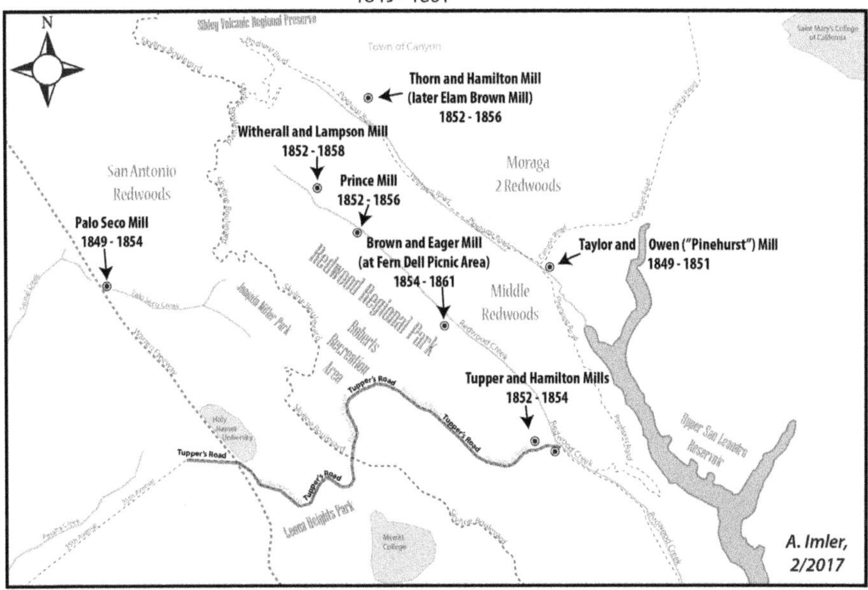

Locations of the steam sawmills that turned the East Bay old-growth redwood forest into lumber. *Courtesy Alan Imler Cartography.*

The Palo Seco Mill is believed to have been located on the creek of the same name, at its confluence with Cinderella Creek.[92] The original mill was powered by water flow. Steam capability may have been added later. Since nails were a scarce and costly commodity, the mill was reportedly held together with wooden pegs.[93]

Based on historical records and the creek locations, the mill site is believed to be on today's Joaquin Miller Court, just uphill from Mountain Boulevard, a few hundred feet within the park gate.[94]

The location was convenient. It was downslope from the tall San Antonio redwoods. It had the advantage of maximum seasonal water flows in the conjoined creeks. And located as it was at the head of Dimond Canyon, logs or lumber could be trucked along Sausal Creek, down the grade that is today's Park Boulevard and down the ridge along Fourteenth Avenue to the *embarcadero* on San Leandro Creek.

Henry Meiggs bought the mill from Joseph Lavigne and Jean Baptiste Bajoux before construction had been completed.[95] By 1849, Meiggs, a San Francisco lumber tycoon, had sent "a small army of loggers" into the woods.[96]

About fifty men were employed at the mill. It supplied the lumber to build the Moses Chase home on East Tenth Street, reportedly the first frame house to be built in Oakland.[97]

Later, Meiggs sold the mill to banker Volney Moody, a founder of the city of Piedmont. By the fall of 1852, Daniel A. Plummer had purchased the mill.

Operations did not end in 1854, when the surrounding forest had all been cut down. While it took only five years to reduce the San Antonio redwoods to stumps, the Palo Seco Mill remained in operation until at least 1911, making wood products of pine and eucalyptus.

Sawmills in the Middle Redwoods

The Prince Mill (1852–56)

The first of four major sawmills in the Middle Redwoods was built and operated by two brothers, Thomas B. and William Prince. It was located at the intersection of today's Stream Trail and Tres Sendas Trail[98] in Redwood Regional Park.

As lumber prices rose in San Francisco in the spring of 1852, Thomas Prince hastened to the forest along Redwood Creek. According to state law passed in 1850, homesteaders were provided with a pathway to purchase unsurveyed lands through a system of school land warrants. The claimant was required to report the lands he wished to acquire and file a survey in order to obtain title.[99] Accordingly, Thomas Prince obtained land title to 320 acres in July 1852 using school land warrants. He sold the land to his brother, William, two years later. The acreage was in a prime location, extending across the canyon from the East Ridge to the West Ridge.

Doing business under the name T.B. Prince and Company, the brothers sold their forest products throughout the East Bay for four years. By 1856, the trees surrounding the Prince Mill were gone.[100]

"Prince's Mill" was such a landmark that when Alameda County was partitioned away from Contra Costa County in 1853, the legal description of the boundary line referred to "Prince's Mill." But a transcription error gave the name as "Princess": "Thence westwardly along the middle of said ridge crossing the gulch one-half mile below Princess Mill; thence to and running upon the dividing ridge between the Redwoods known as San Antonio and Princess Woods; thence along the top of said ridge to the head of the gulch or creek that divides the ranchos of the Peralta from those known as the San Pablo Ranchos (i.e. Wildcat Creek)."[101]

The Witherall and Lampson Mill (1852–58)

The second mill in the Middle Redwoods was a short distance northwest of the Prince Mill, in Tres Sendas Canyon. Joseph Witherall and Nathaniel Lampson built their mill on land claimed by Thomas Prince. Despite his protests, Witherall continued logging. Prince sued. In January 1853, Prince won a court judgment of "forcible entry" against Witherall.

Thereafter, the men promptly agreed that Prince would lease the land to Witherall. Logging continued. To settle their debts, Witherall and Lampson were forced to sell the mill to Luther Mills and James Vantine, who in turn sold it to Henry Spicer.[102] Spicer continued to operate the mill until the forest was exhausted.

Tupper and Hamilton Mill (1852–54)

Chester Tupper and Richard Hamilton arrived in the Middle Redwoods late in 1852. They illegally claimed 360 acres of land and set to logging. For about eighteen months, the operation did well, supplying wood for construction of East Bay towns. But lumber prices dropped in 1854—the same year the mill was destroyed, probably by fire.

Tupper and Hamilton found themselves deeply in debt. The sheriff distributed their assets—in the form of carts, oxen and cut lumber—to their creditors. The mill was never rebuilt.

In the 1977 resource analysis of Redwood Regional Park by the East Bay Regional Park District, the location of the third mill is identified precisely as the Fern Dell picnic area.[103]

However, Sherwood Burgess in 1951 described the location as "near the mouth of Redwood Canyon, where the road now enters the park." The location of the park entrance has shifted during the past decades.

Brown and Eager Mill (1854–61)

Thomas Eager, who had previously logged redwoods in Santa Cruz, staked a claim for 160 acres of public land about one mile north of the Tupper and Hamilton site.[104] At the same time, Erasmus Brown, who had done business with Eager before, staked his own claim in the same area. In 1854, the Mormon men agreed to go into partnership.

They built a steam sawmill for $10,000 just south of the boundary line between Contra Costa and the newly incorporated Alameda County.[105] Two years later, in 1856, Brown purchased Tupper's land after his mill was destroyed. The new supply of timber allowed Brown and Eager to continue milling. During its period of operation, the Brown and Eager Mill was the most prosperous operation in Mill Canyon.

In November 1857, Daniel Plummer bought out Thomas Eager's interest for $3,500. Plummer owned the Palo Seco Mill but had cut all the redwoods on the west side of the hills. It proved to be a bad investment for Plummer; the Middle Redwoods were also logged out by 1858. Litigation ensued.[106] In 1861, Brown and Plummer sold their stump land, with dubious title, for $400.[107]

Lumber Production

Logging methods of the 1840s were slow and labor-intensive. But after steam sawmills were introduced, the redwood forest disappeared quickly. Today's Stream Trail area of Redwood Regional Park was then called Mill Creek, or Sawmill Creek.[108]

"Altogether, [the East Bay steam sawmills] produced about twelve thousand [board] feet of lumber per day, which was hauled to LaRue's Wharf at the foot of Commerce Street [Fourteenth Avenue]....It sold for $25–$35 per thousand, most being taken to San Francisco to erect her earliest buildings."[109]

The Moraga Redwoods

The Taylor and Owen "Pinehurst" Mill (1849–51)

William Taylor[110] and James Owen were first to build a sawmill in the Moraga Redwoods, in late 1849. The mill is believed to have been located at the site of the later town of Pinehurst, on the Sacramento Northern Railroad line. In the present day, this is where the two branches of Pinehurst Road make a "T" with Canyon Road, Moraga, in the East Bay Municipal Utility District watershed.

The Thorn and Hamilton (later Elam Brown) Mill (1852–56)

Ignoring Joaquín Moraga's land grant claim, Hiram Thorn and William Hamilton each claimed 160 acres near the site where Taylor's Mill had stood. In 1852, they invested $20,000 in constructing a sawmill.

Elam Brown was acquiring a proprietary interest in Moraga's *Rancho Laguna de los Palos Colorados*, extending eastward to Las Trampas Creek. Brown bought the Thorn and Hamilton land from Joaquín Moraga in February 1853. Brown then ordered Thorn and Hamilton to leave the property.

Negotiations followed. The men agreed that Thorn and Hamilton would purchase the land from Brown if the courts validated Moraga's claim. Meanwhile, they would pay rent to Brown, in the form of sixty dollars per month in cut lumber. If the courts were to reject Moraga's claim, Brown would refund the rent to Thorn and Hamilton. "Long after the last redwood tree had fallen, the courts recognized the area as *Rancho Laguna de los Palos Colorados*, Tract #2," Burgess noted.

The Moses Davis Mill

One other mill is believed to have operated in the Moraga redwoods. Little is known about the Moses Davis Mill, but it is cited by Frederick Monteagle.[111]

LOGGING ROADS TO EAST BAY TOWNS AND WHARFS

Unlike the coastal logging industry in Santa Cruz or Humboldt Counties and points north, loggers could not depend on waterways or oceangoing schooners to carry their logs to customers. Thus, it was the mill owners who constructed some of the first roads in the East Bay hills.

Lumber from the Palo Seco Mill was the most accessible and, thus, the first to be obtained by the village of San Antonio. Tracks for ox carts and skid roads converged at the top of Dimond Canyon on Sausal Creek.

1850s Logging Roads

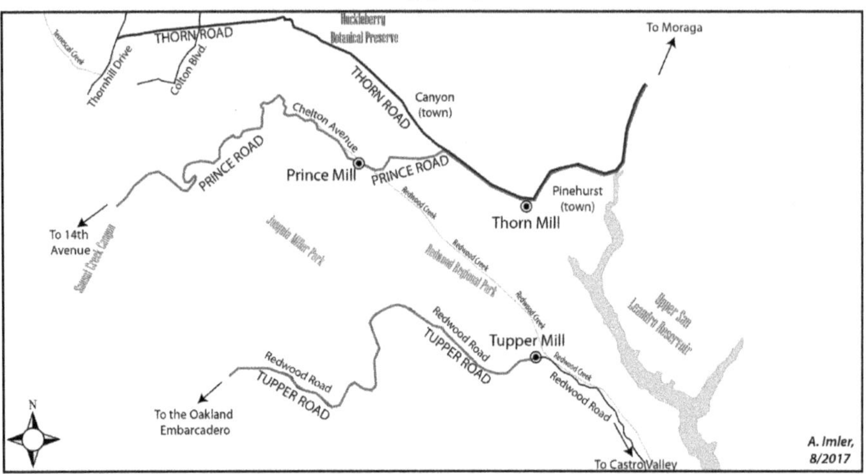

Logging roads built by Hiram Thorn, William Prince and Chester Tupper to bring trees to the waterfront have become thoroughfares in modern Oakland. *Courtesy Alan Imler Cartography.*

Writing a century later, Sherwood Burgess described the road built in 1849 to connect the Palo Seco Mill to the San Antonio waterfront: "After passing through the canyon, the road ascended the ridge leading directly to the embarcadero, along the road followed today by Park Boulevard and Thirteenth Avenue."

In 1852, before Alameda County was incorporated, Contra Costa County built a road from the Prince Mill into Castro Valley. Its approximate route followed the Stream Trail to today's Redwood Road, continuing eastward along San Leandro Creek.

The same year, Oakland lumberyard operators sought a direct route to the redwoods. Lake Merritt had not yet been bridged at Twelfth Street; it was an immense tidal basin that teamsters needed to circumvent. The new county road followed Broadway northward, ascending the hills along Broadway Terrace to Mountain Boulevard, where it met Prince's Road near today's Chelton Drive.[112]

In 1856, the county built a road from Thorn's Mill, following San Leandro Creek into Castro Valley.[113]

Prince's Road

In 1852, as a condition of the Prince brothers' land grant, the county ordered them to survey a public road from Lafayette to the San Antonio Redwoods, continuing across the hills to Oakland. The closest road that then crossed the hills was to the south at Mission Pass, today's Sunol grade.

William Prince and Thomas Bradley did the survey, and Contra Costa County built the road. It started in Lafayette and followed approximately the present Moraga Road and Canyon Road in Moraga, to the Pinehurst mill site. It followed today's Pinehurst Road up the creek below the town of Canyon, at which point it climbed the hill to the East Ridge.[114]

Although it would have been a more direct route to follow today's East Ridge Trail to the Skyline Gate, Prince's Road dropped down into the canyon near the brothers' mill. (A section of the 1852 road—called the Prince Trail from the East Ridge to the Prince Mill site—is still in use.)

To connect the Prince Mill to Castro Valley, the road followed Redwood Creek out of the canyon to the vicinity of today's Redwood Road, where it joined the earlier county road along San Leandro Creek. To connect the Prince Mill with Oakland, the road continued along the route of today's Mill Trail up the canyon to the skyline.

Burgess continued the description of Prince's Road: "After running a short distance south, the route followed the nose of a ridge near present day Chelton Drive to the head of Dimond Canyon, near the present intersection of Park and Mountain Boulevards."[115]

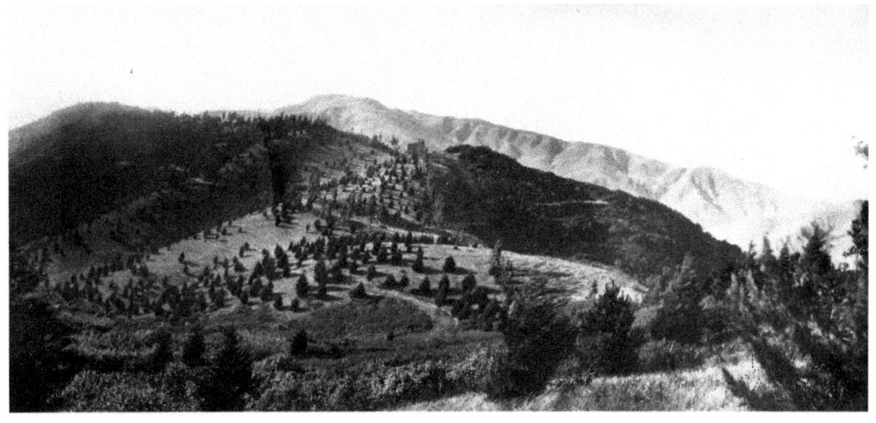

When the Havens plantation trees were still small, the contours of Shepherd Canyon were visible—along with a segment of "the old logging road," primarily built by William Prince. *Courtesy East Bay Regional Park archives.*

According to the 1977 EBRPD Redwood Regional Park Resource Analysis, however, Prince's Road turned south at today's Chelton Drive and descended into Shepherd Canyon. "After the logging era, [Prince's Road] was abandoned and forgotten, but the grade is still there," the EBRPD document concluded.

In a 1941 interview, Oakland dentist John Engs [116] reminisced about riding his horse through the Oakland hills in the years following the San Francisco earthquake of 1906. Engs described "part of an old logging road behind the Dimond Scout Camp" that connected to Joaquin Miller Road.

The Dimond Scout Camp once stood on the site of today's Joaquin Miller Elementary and Montera Middle Schools. Scout Road, upslope from the schools, retains the historic name. Dr. Engs thereby suggested that Prince's Road followed today's Scout Road and Mountain Boulevard southeastward, past the Palo Seco Mill site, to Joaquin Miller Road.

While the Prince Mill was in operation, today's Tres Sendas Trail in Redwood Regional Park was developed as a skid road, over which oxen dragged logs.

Tupper's Road

Tupper and Hamilton wanted a shorter and more direct route from their mill to San Antonio than Prince's Road. In 1853, they built a road following Redwood Creek up to the skyline ridge. There they built a second crossing—today's Redwood Road.[117] As the city of Oakland grew, the flatland section of the road was named Thirty-Fifth Avenue.

Whereas the Prince and Thorn Roads accessed Oakland and San Antonio through Sausal Creek Canyon (Park Boulevard), the Tupper road allowed logs to be carted to the *embarcadero*.

Shortly after Redwood Road was completed, the county extended High Street from the city of Alameda northeast to the base of the Oakland hills, meeting Tupper's Road. Along the way, it passed the Thompson Artesian Water Company works.[118]

Thorn's Road

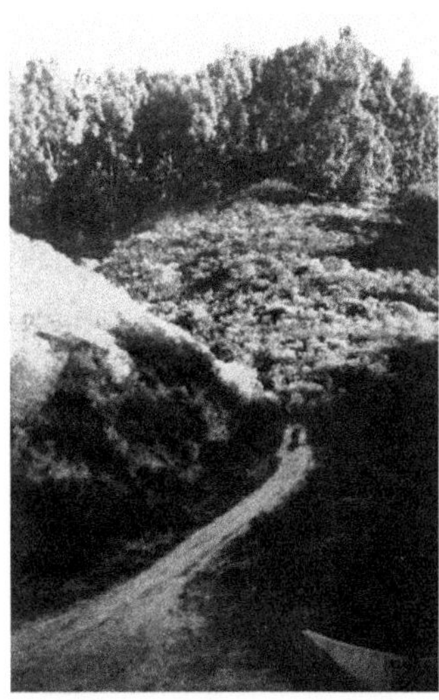

Hiram Thorn regained his fortune by operating his toll road between Oakland and Moraga. This view would appear to be looking eastward up the road that would become today's Thornhill Drive. *East Bay Regional Park archives.*

Hiram Thorn, too, was dissatisfied with Prince's Road. In 1853, he formed the Thorn Turnpike Road Company, capitalizing it initially with $5,000 of his own. He also recruited some local investors.

The road Thorn built started at his mill in the Moraga Redwoods, near the town of Pinehurst. Thorn's Road followed the San Leandro Creek watershed up to the ridge overlooking Oakland to the west. It crossed the saddle at the north end of today's Huckleberry Botanical Preserve.[119]

Thorn's Road then dropped down today's Thornhill Canyon,[120] continuing south approximately along the route of Mountain Boulevard to a point near the top of Dimond Canyon. There it met the "old road to San Antonio," today's Park Boulevard, and the "new road to Oakland," Broadway Terrace.

In 1854, the price of lumber dropped. One of Thorn's investors sued him. In 1855, during an economic downturn, other investors withheld funds that they had previously offered. With his businesses insolvent, Thorn appealed to the county supervisors. They agreed to let him complete his project and operate it as a toll road. After a few years of doing so, Thorn regained his fortune.

In the 1953 survey description that defined the boundary features of the new Alameda County, Thorn Road was used as a landmark to locate the township of Brooklyn:

> *Commencing at the center of the Thorn Road (the same being the road leading from East Oakland to Moraga Valley), where the same crosses the line dividing the counties of Alameda and Contra Costa, on the summit of the mountains being also the easterly corner of the Oakland Township; thence southwesterly along the center of said road to the head of Indian Gulch;*[121] *then down said gulch to the north branch of the estuary of San Antonio (the same being now known as Lake Merritt).*[122]

Unlike Prince's Road, of which most sections were abandoned, the Thorn Road continued to be used well into the twentieth century. During the era of the Sacramento Northern Railway, today's Pinehurst Road was built over the Thorn Road, along the railroad grade past the town of Canyon. This road crosses the hill crest into Shepherd Canyon today.

HOW EAST BAY CREEKS WERE USED IN THE 1850S LOGGING ERA

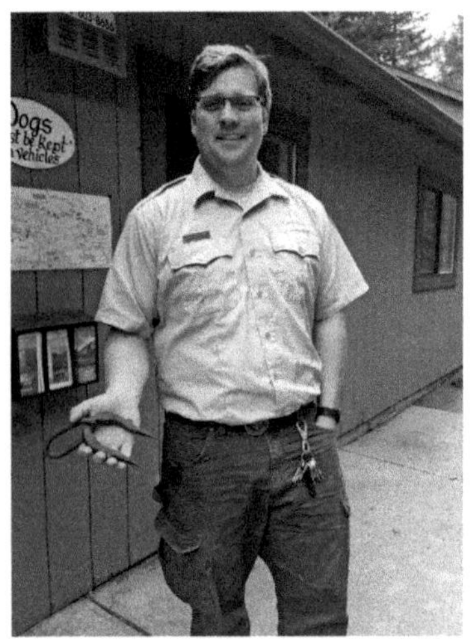

Redwood Regional Park supervisor Justin Neville displays a rusted clamp dating from the logging days. The park district catalogues "field finds" in its archives. *Amelia S. Marshall photo.*

Local historian Dennis Evanosky has studied the area near the probable site of the Palo Seco Mill. "The 1850s loggers used Palo Seco Creek to get the logs down the hill to the mill," Evanosky explained. "The low-hanging branches were called 'slice.' The pine needles and twigs were called 'duff.' They pushed the slice and duff into the creek to make a single-use path for the ox teams, to get the logs to the mill."

But were the San Antonio redwood trees "floated down the creeks," as some sources maintain? Forester John Nicoles is skeptical:

> *The notion that you could float anything resembling a log down one of our creeks is not believable. There would be no way to float an eight-foot length of straight lumber down one of our East Bay creeks. Shingle bolts or firewood during the wettest season—maybe. But dry periods have been* [a standard feature of the Bay Area climate] *for the past two hundred years.*
>
> *When we consider the creeks of the East Bay, the original stream channels looked very different from how they are today.*
>
> *According to Burgess, in Redwood Regional Park there was an initial period of logging pre-Gold Rush, and by 1855 it was all over. The stream*

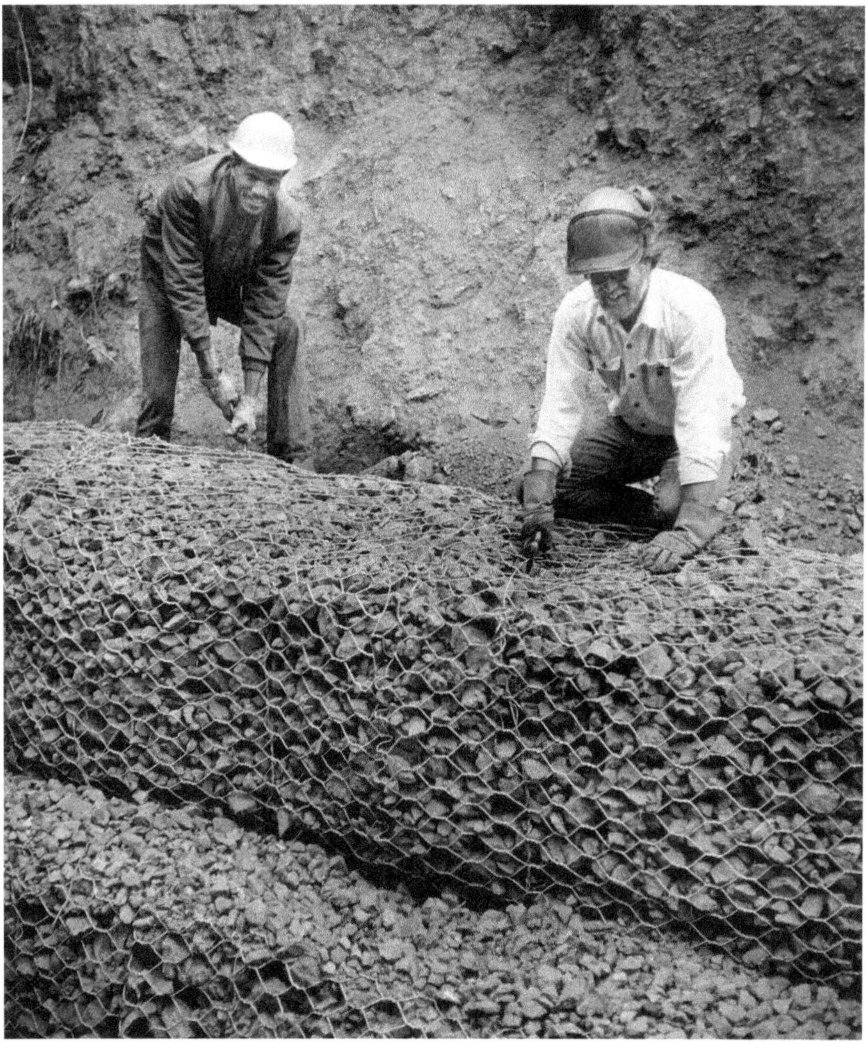

Terrence Price and John Nicoles stabilize the banks of Redwood Creek after a rainy season, circa 1981. *Courtesy East Bay Regional Park District archives.*

channels we see today have been affected by runoff from all the paving and development that has followed.

From the forestry point of view, a soil that will support a forest can absorb as much water as you can throw at it.

But now, instead of a steady flow into the stream from saturated ground throughout the season, there is a higher peak flow after it rains. This has caused the stream channels of today to be much more deeply incised.[123]

Powering the Early Mills

How were the early mills powered? To turn the saw blades, did the operators use steam power or hydro power, where volumes of moving water turned the rotors? "Good question," Nicoles replied.

> *The factor that is most important for using hydro power is not the head, or elevation from which the water falls to provide energy to turn the wheel, but the volume of water available. We have always had a limited amount of water in the East Bay redwoods, with water shortages being a frequent occurrence. It is likely that the mills were operated seasonally.*
>
> *I learned something interesting from Ralph Sturgeon, the son of the original owner of Sturgeon's Mill, located near Sebastapol. In the old days, they would log for two or three days, then they would fire up the mill and run for two or three days. Nowadays we think in terms of mills running 24/7, but that was not the way it was back then. If you're not running on a daily basis, you can impound water to power the mill. The secret to powering the mills may have been intermittent operation.*

It is likely that the most productive of the mills—Prince's, Thorn's, Tupper and Hamilton's and so on—were all steam powered. "Once steam-powered mills were introduced, the redwoods were cut down very quickly," observed Dee Rosario, supervisor of Redwood Park from 1998 to 2013.

Most historical documents describe the Palo Seco Mill, which processed the entire west-facing old-growth redwood forest, as having been steam powered. Again, John Nicoles is skeptical: "The Palo Seco Mill was supposedly operated by steam. I have carefully examined the historic photos of the Palo Seco Mill, but I see no evidence that it was steam powered."

Many questions remain unanswered about the lumbering operations that clear-cut the original East Bay redwoods.

Where Did the Lumber Go?: Earliest Commercial Logging in the East Bay

In the earliest days of commercial logging in the redwoods, the lumber was used in the East Bay. Burgess cited a load of lumber shipped across Suisun

Bay in 1848 to Dr. Robert Semple, a founder of the city of Benicia. The following year, lumber was used in building downtown Martinez. What remained of the timber harvest was used on nearby farms and ranches.

Some lumber was sent to San Francisco to rebuild the city after the series of major fires between 1849 and 1851. But the supply from the East Bay was limited. The price of lumber in San Francisco dropped significantly as international lumber imports increased. In January 1850, lumber prices were quoted at $150 to $300 per thousand board feet. By December 1850, the price had dropped to $50 per thousand board feet. Redwood planks were being sold at $40 per thousand board feet. In 1852, the price rose to $100 per thousand board feet as lumber shipments from the East Coast declined.[124]

As late as October 21, 1853, a report in the San Francisco newspaper *Alta California* listed the sources of recent lumber shipments. No wood from Contra Costa was mentioned.

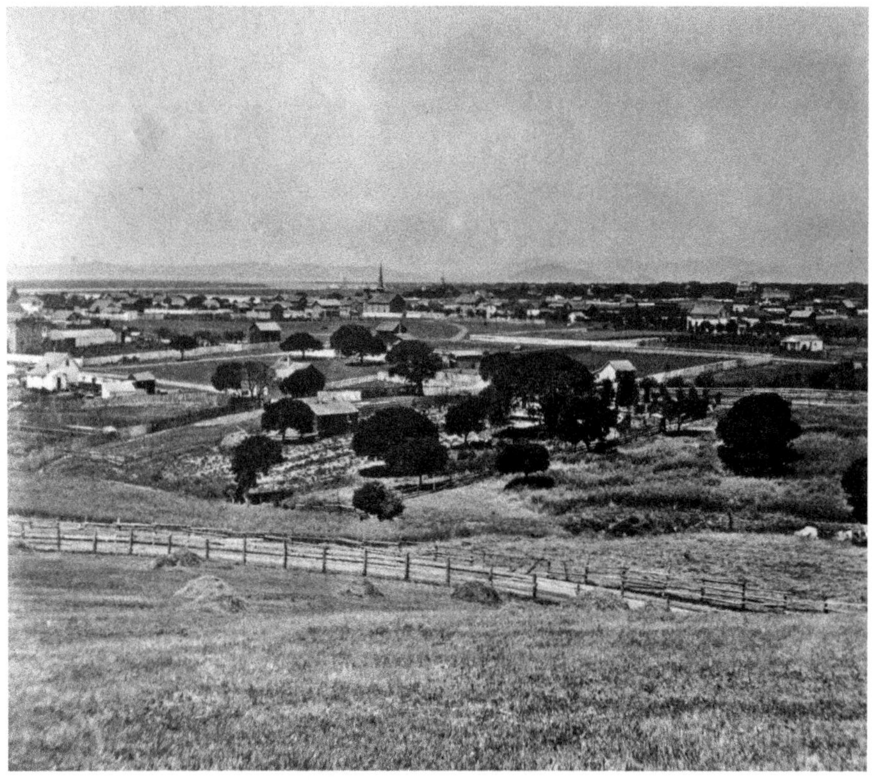

Lumber from the redwoods was used to build San Antonio. The fence line in the foreground is believed to have been the boundary of Henderson Luelling's farm. *Courtesy Society of California Pioneers.*

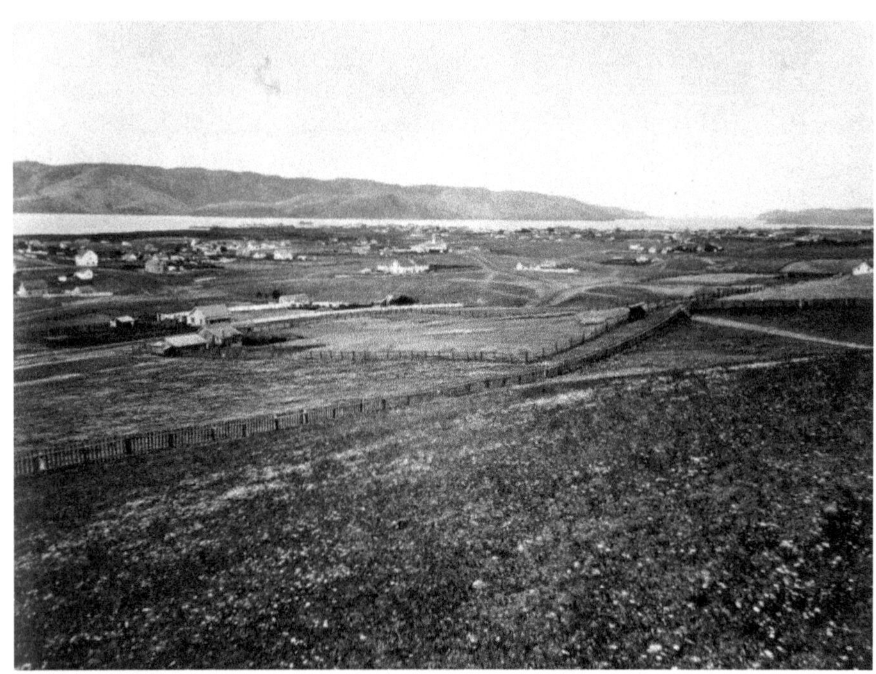

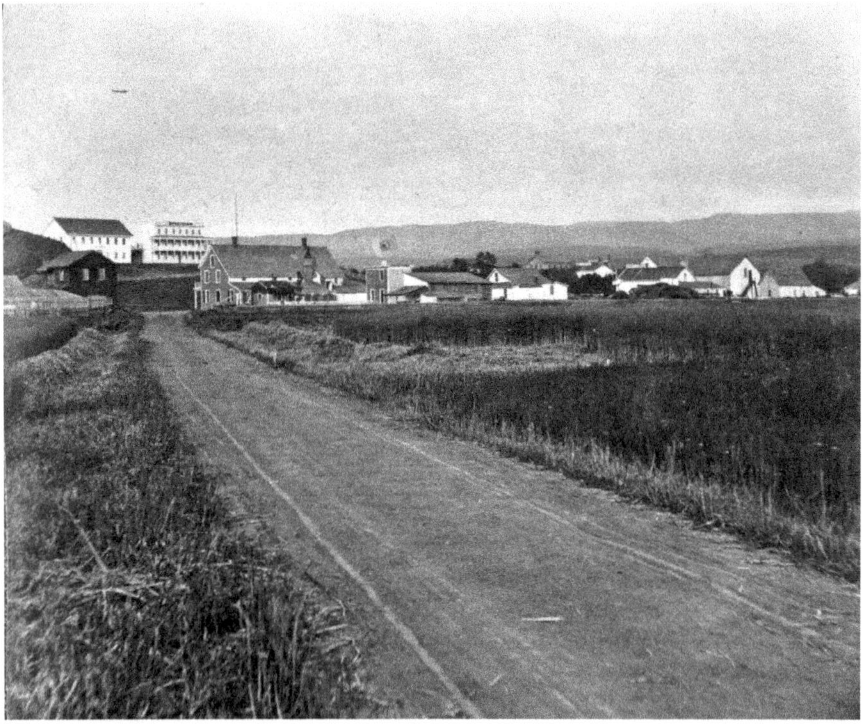

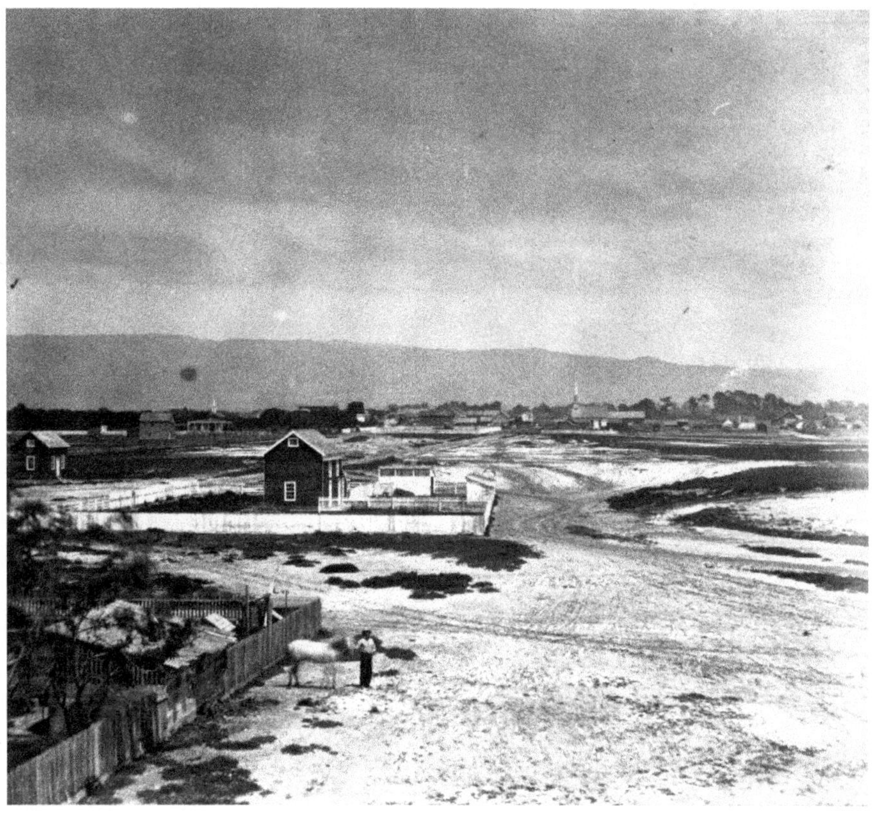

Above: East Bay hills redwood lumber built Alvarado. Boasting a productive field of artesian wells, Alvarado became the first "seat of justice" of Alameda County. *Courtesy Society of California Pioneers.*

Opposite, top: Redwood lumber from the East Bay hills was shipped up the Carquinez Strait to build the village of Benicia. *Courtesy Society of California Pioneers.*

Opposite, bottom: Redwood lumber from the East Bay hills was used to build the city of Haywards to the south. *Courtesy Society of California Pioneers.*

After production was increased through the introduction of steam sawmills, lumber was carted to local towns, including Martinez and Alvarado. Burgess quoted a report on November 7, 1854, in the *Sacramento Daily Union* announcing the arrival of a Contra Costa lumber shipment. By then, the towns of the East Bay were experiencing rapid growth. Oakland, Clinton, San Antonio, Alameda, Alvarado and Haywards (today's Hayward) were buying redwood lumber as fast as the sawmills could deliver it.

After the 1906 Earthquake

Many writers have suggested that San Francisco was rebuilt with redwood lumber from the East Bay after the devastating earthquake and fires of 1906, but for the most part, this was not the case.

Forester John Nicoles maintained that after the East Bay redwood forest had been clear-cut, its second-growth trees were still not large enough for substantial building material: "By the time of the 1906 San Francisco earthquake, most of the redwood lumbering industry was centered on the North Coast. There, logs could be floated down rivers and carried to San Francisco on ships. Much of San Francisco was rebuilt with Douglas fir rather than redwood."

THE EAST BAY REDWOOD FOREST TODAY

Second Growth or Third?

Twenty-first-century observers of the East Bay Redwood forest often refer to it as "second growth." How do we know which trees are actually third growth, their progenitors having been cut down after the 1850s logging boom?

It is likely that any logging of second-growth trees would have been done in the San Antonio redwoods, on the west-facing slopes in today's Joaquin Miller Park, as early as the 1870s. There is some indication that the Palo Seco Mill continued in operation after the mills in the Middle and Moraga Redwoods were abandoned. "It is possible that after the old-growth forest was cut over, there was a period of cutting second growth in the 1870s. Redwood round stock—twenty- or thirty-year-old stems—could have been used for posts. I've seen some like that in coal mines east of Mount Diablo," said Nicoles.

The scientific way to determine the ages of the trees in either the Stream Trail area or on the west-facing slopes requires taking core samples to count the number of rings. Dee Rosario suggests that a less precise and less invasive approach is simply measuring trunk diameter. "Most second-growth trees were cut between today's Roberts Park and Chabot Space and Science Center," Rosario noted. "You can tell by the size of the trunks."

Why Giant Old-Growth Stumps Were Never Found

Park rangers and naturalists, as well as interested visitors, make a practice of searching for the widest redwood fairy rings and standing trunks in the East Bay hills.

In the decades after Gibbons brought visitors to picnic in his thirty-three-foot stump, a surprising dearth of similar stumps were in evidence.[125] Why? Gibbons himself provided the answer, noting that five hundred or more shoots would propagate from the largest stumps:

> *Such extreme tenacity of life and capability of propagation would ensure its perpetuity in this isolated location but for the fact that settlers in the region are continually working up the roots for fuel....For years past, squatters of foreign birth have occupied the canyons and derived a revenue by cutting off the saplings and digging out the stumps to convert them into firewood.*[126]

Old Survivor

Paul Covel, longtime naturalist for the Oakland Parks Department, is credited with identifying in 1969 the sole surviving old-growth redwood tree in the East Bay hills. By then, Gibbons, Burgess and others had concluded that not a single tree had been left behind by the rapacious nineteenth-century loggers.

"Old Survivor," also known as "Grandpa," had two things going for it. First, although it stands very close to the streets of Oakland, getting up to the rocky, brush-covered outcropping into which it is rooted is a challenging trek for all but the most agile climber. Second, the gnarled and twisted shape of the tree likely made it less attractive as lumber.

"Old Survivor" grows in Leona Heights Park,[127] above the Leona Lodge, where it can be approached via trails and fire roads, including remnant sections of the old York Trail. This open space area is riddled with old mines. The infamous "Devil's Punchbowl," a deep shaft from which minerals (including rhyolite, sulfur and possibly also hematite) were excavated was once nearby. However, after several accidents and much unlawful behavior there, the Punchbowl was filled and graded away at the time Merritt College was built in the early 1970s.

Covel discovered "Old Survivor" around the time of extensive construction activity in the hills. In addition to the construction of

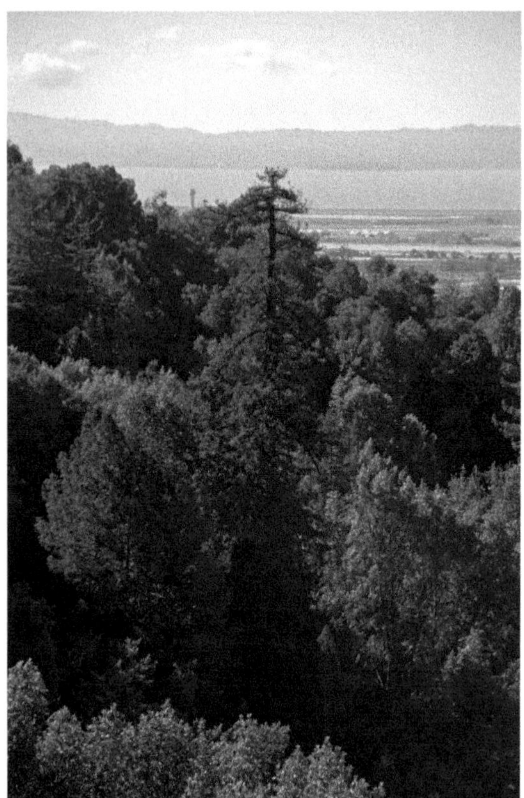

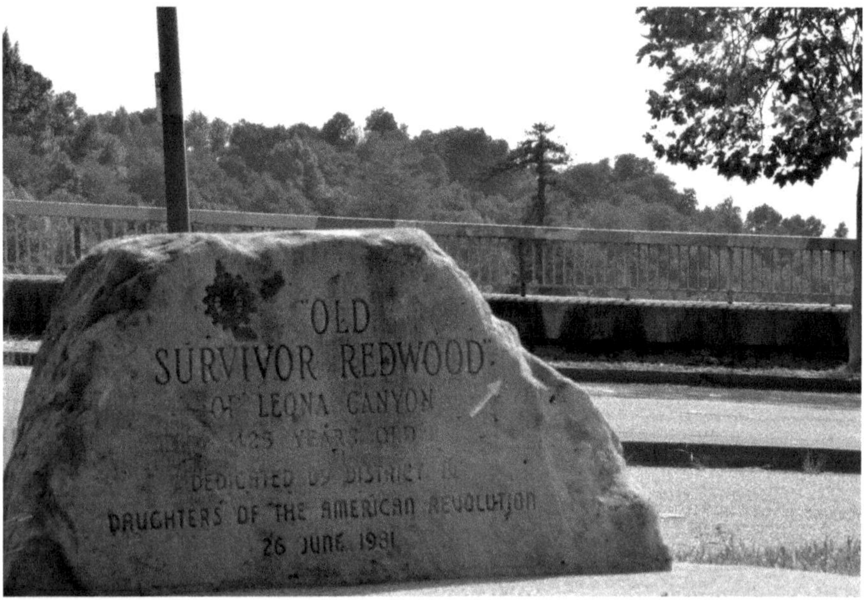

This page: "Old Survivor," believed to the only first-growth redwood that remains from the 1850s logging era, can be viewed in Leona Canyon from the front of the Carl Munck School. *Bill Imler photo.*

Merritt College and Skyline High School, Redwood Road was widened to accommodate increased traffic between new housing in the hills and the Fruitvale BART station.

To confirm the value of his discovery, Covel enlisted Glen Strouse of the Humboldt State College forestry department. Strouse took a tiny core sample and then used microscopic techniques to determine that the tree was (in 1969) between 415 and 420 years old.[128]

Ambitious hikers wishing to visit "Old Survivor" can take the fire road between the Leona Lodge at 4444 Mountain Boulevard and the Campus Drive overpass. Perhaps the best vantage point, however, is from the front parking lot of the Carl Munck School at 11900 Campus Drive. An interpretive plaque shows where to look for the spire of "Old Survivor."

Chapter 6
SETTLERS

And now the greedy, blue-eyed Saxon came.
—Don Antonio Maria Peralta

Founders of Oakland: Carpentier, Adams and Moon

The first Yankee settler in what was to become the city of Oakland was Moses Chase. In the winter of 1849–50, Chase pitched his tent at what is now Jack London Square and promptly commenced hunting for food. The Patten brothers joined Chase in February 1850. Next to arrive were Colonel Henry S. Fitch and Colonel Whitney. In May 1850, a trio arrived who would permanently shape the history of the East Bay: Edson Adams, A.J. Moon and Horace Carpentier.[129]

The three built a shanty at the foot of Broadway, near the wharf used by the Peralta family. When Antonio Peralta asked them to leave his property, the Yankees refused, maintaining that the land belonged to the United States government. Peralta obtained a writ of ejectment from the county court in Martinez, and "a posse of men, under Deputy Sheriff Kelly, was sent out to eject them." The squatters, however, refused to leave. Antonio Peralta, evidently seeking to make the best of a bad situation, agreed to give Carpentier and his partners a lease for several acres.

Carpentier, who spoke fluent Spanish, ingratiated himself with Jose Domingo Peralta and other land grant heirs. The lawyer accepted their

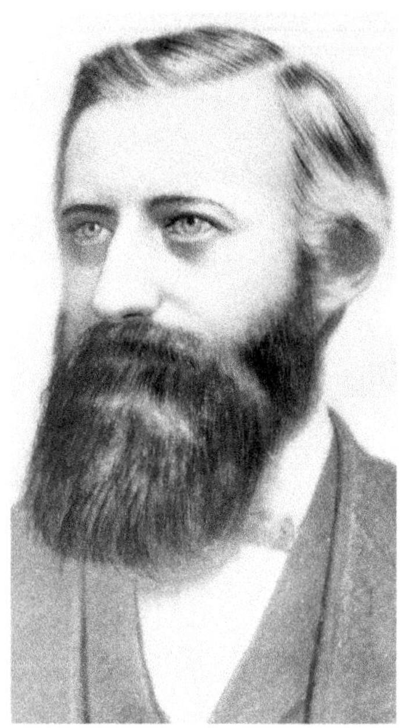

Right: Horace Carpentier was a recent Harvard law school graduate when he arrived in Oakland. He accepted the *Californios*' hospitality and then defrauded them of their land. He was elected the first mayor of Oakland with more votes than there were residents. *Courtesy Oakland Main Library, Oakland History Room.*

Below: Lumber from the East Bay redwoods arrived at the foot of Broadway to be loaded onto ships. *Courtesy Society of California Pioneers.*

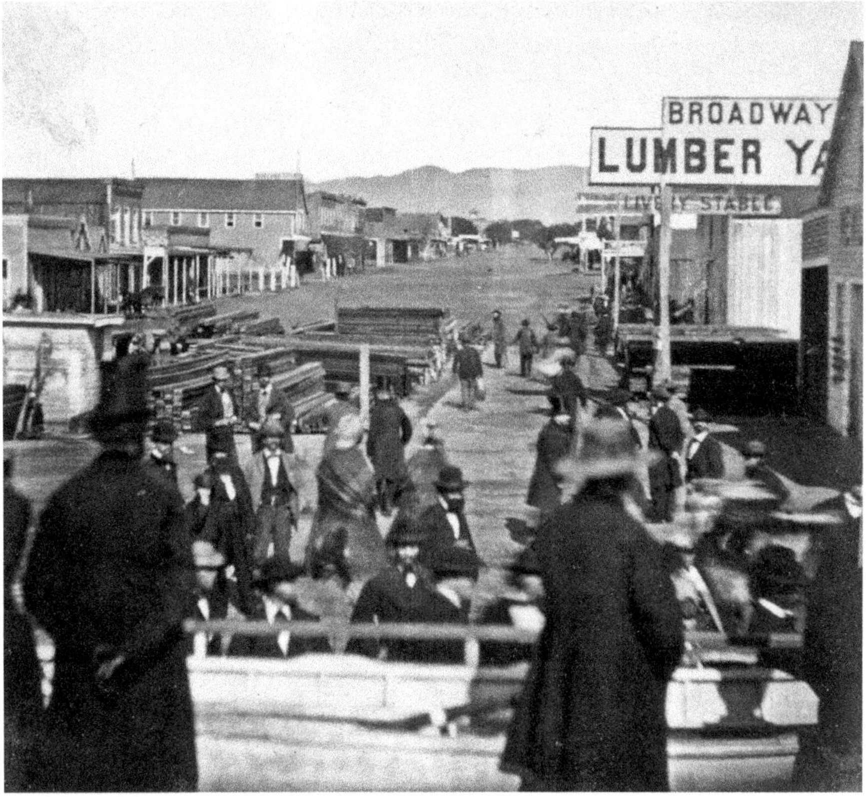

hospitality in order to gain title to their real estate. Finally, nothing was left of Domingo's land except for his house. In 1865, he died.

Carpentier also wanted to possess the Moraga redwood forest. Through a complex series of legal maneuvers, the lawyer acquired most of the Moraga and Bernal land.

Carpentier, the Politician

By 1852, Carpentier had obtained an official position in the state legislature, which met in Benicia, then the state capital.[130] On May 4, 1852, Carpentier succeeded in getting an official act of incorporation passed to create the town of Oakland on the site of Contra Costa village. The population was then nearly two thousand.

Carpentier was elected Oakland's first mayor in 1854, having received more votes than the number of residents.[131] Less than one year later, he was driven out of town by an angry mob after it became known that he had, through trickery, secured the rights to control the entire waterfront.

The original village of Oakland was located entirely on the west side of the tidal marsh that would become Lake Merritt. To the east were the villages of San Antonio, Clinton and Lynn. In 1856, these were consolidated as Brooklyn Township.[132] Farther east, where dairy cattle would be pastured and fruit orchards planted, were the towns of Melrose and Fairfax. Within a few decades, these would all be within the Oakland city limits.

Settlers of Fruit Vale and the Sausal Creek Watershed

From an aerial view above the East Bay, Sausal Creek cuts a green pathway from its headwaters below Redwood Peak[133] downhill to the bay. At the time when the private water companies went into business, the creek met the bay among willow groves[134] near today's Fruitvale Bridge.

Today, Sausal Creek remains the major East Bay creek whose course is least modified by urbanization. While many sections of Sausal Creek have been piped through concrete culverts, about half its length remains wild and open to daylight.[135] Other major East Bay streams—San Leandro Creek, Arroyo Viejo/Elmhurst, the Peralta/Courtland/Seminary system and Temescal Creek—have not fared as well.

Above the Sausal Creek Canyon, its tributaries are Palo Seco and Cinderella Creeks in today's Joaquin Miller Park; Cobbledick Creek, which parallels Scout Road; and Shepherd Creek. Below the base of the hills, the gorge widens. The creek continues downhill into Dimond Park. When only the Ohlone people were present, redwoods grew all the way down to below today's MacArthur Freeway.[136]

Yankee settlers found a clear, cold stream with a thriving population of trout. The waters had carried nutrient-rich soil from the forested slopes down to the flatland from prehistoric times. Not even the logging slash in the upper creek had spoiled its course through the flatlands.

In earlier times, the Peralta family saw the creeks as geographic boundaries across the coastal plains. Today, environmentalist sensibilities again view the Sausal Creek watershed as a unified biosphere.

The gradual development of the Fruit Vale countryside began with grand estates created by hardworking people. Three key players in this were Henderson Luelling, Caspar Hopkins and Hugh Dimond.

Henderson Luelling (1809–1878)

Henderson Luelling[137] had become a well-known horticulturalist by the time he bought land along Sausal Creek for a nursery. He and his first wife, Elizabeth, had been active in the antislavery movement in an Iowa Quaker colony prior to the Civil War. The Luelling house had been a stop on the Underground Railroad.

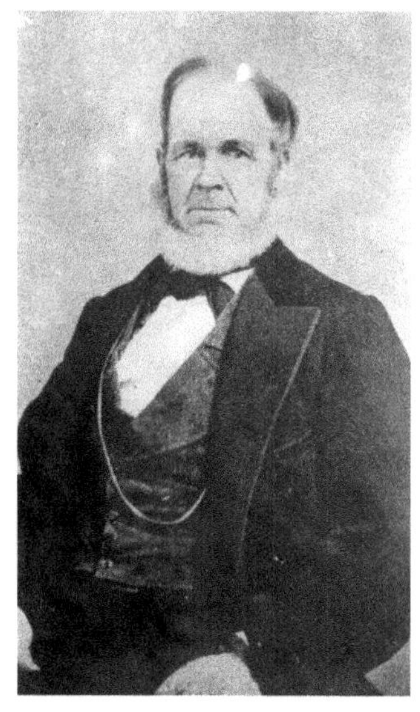

Horticulturalist Henderson Luelling founded the Fruit Vale district of Oakland before fleeing to Central America with a "free love" sect called the Harmonial Brotherhood. *Courtesy Bancroft Library.*

Seeking opportunities in California, Luelling and his family came to East Oakland in 1853, where he created a large orchard on the south bank of Sausal Creek. He planted the first apple and pear trees in California, as well as cherries and other stone fruit. His daughter-in-law, Elizabeth, wife of his son, Alfred, is credited with naming the area Fruit Vale.[138]

The family built "an imposing house near what is now the south entrance to Dimond

Park."[139] The property included the hill on which the historic Altenheim was subsequently built.

Two wives of Luelling had predeceased him, and he had remarried twice by the time he arrived in Oakland. In 1859, he sold his house and most of his land to John B. Weller, governor of California. He surreptitiously left his wife, Mary, and set sail for Honduras with a "free love" society called the Harmonial Brotherhood. Mary was left destitute; her friends built her a small house near their former farm.

Things did not work out well for the free love colony. Luelling returned to Northern California in 1860, where he is believed to have lived under the pseudonym Henry.[140] He died of heart disease in San Jose in 1879.

At Mountain View Cemetery, the grave marker for Luelling and his family identifies him as the "Father of the Pacific Fruit Industry." "The fruit industry ended up bringing greater wealth to the state of California than all of the gold ever produced here," observed Oakland historian Michael Colbruno.[141]

Caspar Hopkins (1826–1893)

Caspar Thomas Hopkins was an original thinker, a prolific writer and a prodigious developer of public works—in the East Bay, second only to Anthony Chabot.

Hopkins graduated in 1847 from the University of Vermont. He then came to California in 1849. Four years later, he married Almira Burtnett. They had four children. After trying out various endeavors, in 1861 Hopkins began building a successful business in marine and fire insurance. His California Insurance Company was the first on the West Coast. He served as president for thirty-five years. Dr. Samuel Merritt was vice-president.

Hopkins moved from San Francisco to Fruit Vale with his family in 1868, seeking "fresh air…and freedom from obnoxious neighbors."[142] He found a nascent community taking shape. While Hopkins's time in Fruit Vale overlapped only one year with that of Henderson Luelling, other neighbors had followed to farm, plant trees and graze dairy cattle.

German immigrants were also forming a rustic enclave. "Forty-eighters," liberals who were forced into exile after a failed attempt to unify the German-speaking provinces of Europe, found the East Bay a welcome place to homestead. Frederick Rhoda arrived first, in 1849. He purchased four hundred acres in Fruit Vale and planted Royal Ann cherries. The Rhoda

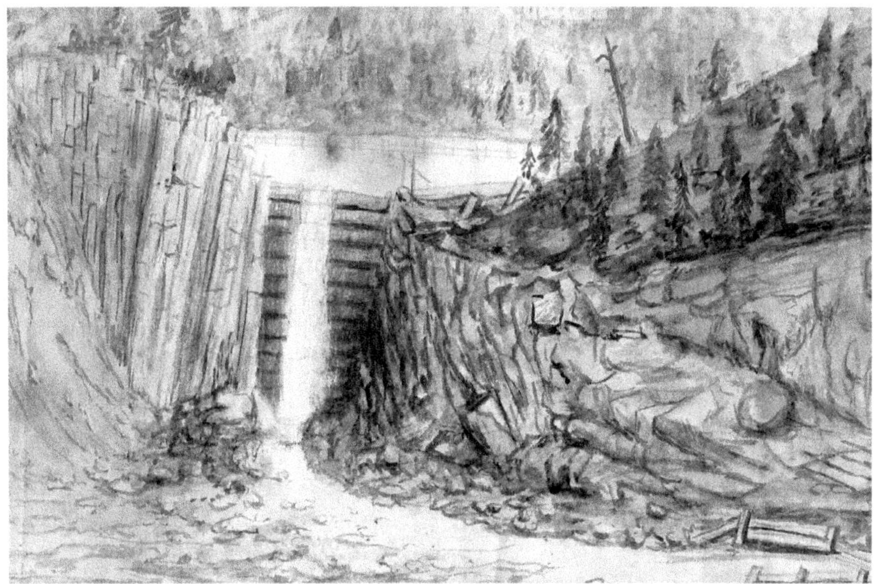

This watercolor drawing, which was archived with a receipt dated 1866, appears to depict Caspar Hopkins's dam on Sausal Creek, from which water was piped four thousand feet downhill to irrigate the orchards of Fruit Vale. *Amelia Sue Marshall Collection.*

House remains as a private residence on Whittle Street. Nearby, on upper Fruitvale Avenue, the Victorian home of the Brendermuhl family, cherry farmers, also stands.

Hopkins devoted his spare time to creating his Alderwood estate, just downslope from the Luelling farm. In his memoirs, he described the home he created with his family:

> *For $6000 I purchased six acres of an apple nursery that had been allowed to grow up, there being no market for the trees....* [He replanted the apple orchard with] *two hundred and fifty fine cherry trees, peaches, almonds, apricots.... There was a small house which I repaired, thinking we could live in it while the children were at school. But it was close quarters.*
>
> *The location was, however, beautiful—in the bottom of the long, narrow valley of Sausal Creek, which penetrated the mountains east of Oakland through a steep, narrow, well-wooded canon.... The place was sheltered from the prevailing southwest winds, and its altitude being 125 feet above sea level, it was rarely visited by fogs.... The creek meandered through*

> *the lot in form like the letter S* [it has since been straightened and spoiled] *and was lined with huge oaks, laurel, alder, and buckeye trees. The large alders…were most numerous, hence we gave the place the name of "Alderwood."…The improvement of this lovely spot was for several years the joy of my life and I was greatly aided therein by the sympathetic and artistic concurrence of my wife.*
>
> *I designed a large, low, Gothic cottage with wide porches on three sides. The old house removed to a new location in a bend of the creek, formed a part of it.…A new street was opened and fenced on the north side, shortening the drive to Oakland from five to three miles.…I bought four more acres across the creek, on the hillside…*
>
> *We widened Fruit Vale Avenue from forty to sixty feet…the neighbors clubbed together and built a water work which cost $20,000 and has since supplied the vale with water in pipes to every house. We again clubbed together and built the Brooklyn and Fruit Vale Horse Railroad across the hills.…I was president both of the water works and the railroad, and did most of the work of organizing and constructing both.*

The new street on the north side of Alderwood provided a more direct route to downtown Oakland. Initially called Hopkins Street, it was renamed MacArthur Boulevard when Oakland was reconfigured in the era of freeway building that followed World War II. Alderwood was located near the southwest corner of today's MacArthur Boulevard and Dimond Avenue. An unassuming hamburger stand and asphalt parking lot are what we find there today. Yet there are also old palm trees that may have grown on the original estate. Nearby, Sausal Creek emerges from under the boulevard to meander downhill from a row of shops and behind apartment buildings before disappearing beneath the I-580 freeway.

Later, in 1878, Hopkins also developed a real estate tract on the land from which the widow and children of Domingo Peralta had been evicted in 1872. Obligingly, Hopkins named the area along Codornices Creek Peralta Park. The City of Berkeley named the main thoroughfare Hopkins Street.

Hugh Dimond (1830–1896)

The third key figure who shaped the urbanization of Fruit Vale, he whose name was chosen for the district, was Hugh Dimond. This native of County Kerry, Ireland, came to California in 1849 and began to operate a general store and wholesale liquor business in the mining district of Mariposa County.

Returning to the Bay Area in 1862 with a substantial fortune, Dimond bought 267 acres, including the Luelling House with its orchards and hay fields, and all of Sausal Creek Canyon. He married Ellen Sullivan Dimond in 1872. The following year, at age forty-one, he retired from business. The 1876 city directory gives his occupation as "farmer." He and Ellen had three children: Nellie, Hugh and Denis.

Having had the old Luelling house refurbished and enlarged, the family moved into it in 1877. Ellen died shortly thereafter.

The Dimond home was described as an imposing two-story white wooden structure surrounded by a spacious one-story veranda. Today's Dimond Avenue was originally the carriage entrance to the house, on the north side of the redwood grove at the south entrance to today's Dimond Park.[143]

The Dimond family diverted water from the creek for their household use. They also built a small dam and dug out a swimming hole at the base of a rocky cliff for their children. This is the present location of the Lions Club swimming pool.

Hugh Dimond died in 1896. The following year, Denis Dimond built a cottage as a playhouse for his children. Adobe bricks from the original Antonio Peralta *hacienda* were incorporated.[144] Evidently, Denis Dimond obtained the adobe bricks from Henry Z. Jones, the real estate developer who was subdividing the land around the Antonio Peralta *hacienda*.

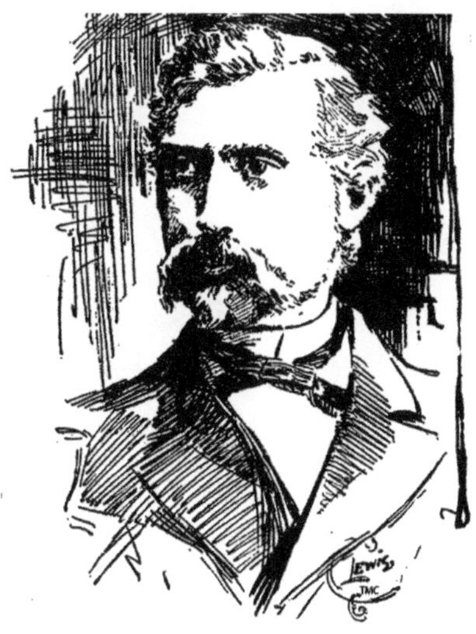

Hugh Dimond, a modest Irish liquor merchant, built his family home on Sausal Creek. The surrounding district was named in his honor. *Courtesy Oakland Main Library, Oakland History Room.*

After the Dimond children had moved away from Oakland, their main house was badly damaged in a fire in 1913. But the adobe cottage remained. In 1917, the Dimond family sold the last twelve acres to the City of Oakland for $24,000.

In the progressive atmosphere of civic pride that followed World War I, the city made an effort to make the new Dimond Park exemplary. Erika Mailman in the May 2, 2000 *Montclarion* quoted *Oakland Parks and Playgrounds* (a book): "The Oakland Park Department set out wide fireplaces to encourage picnicking, and Sausal Creek was so clean that 'breathless children leave their games to quench their thirsts at clear, cool springs.'"

SETTLERS OF THE SAN ANTONIO REDWOODS (ROBERTS REGIONAL PARK)

Frank Havens and the Realty Syndicate

Frank Colton Havens (1848–1918) was largely responsible for transforming the East Bay hills from rolling grasslands to an urban forest, vulnerable to fire danger into the twenty-first century.

Havens was a lawyer from Sag Harbor, New York, who came to California to advance his fortunes through real estate development. Partnering with the colorful Oakland tycoon Francis Marion ("Borax") Smith, of Twenty

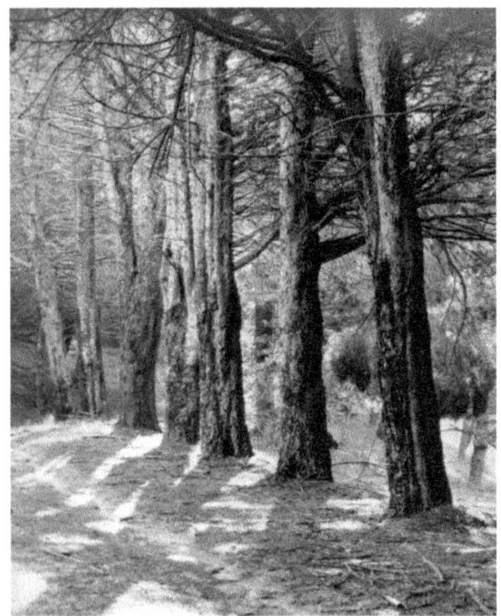

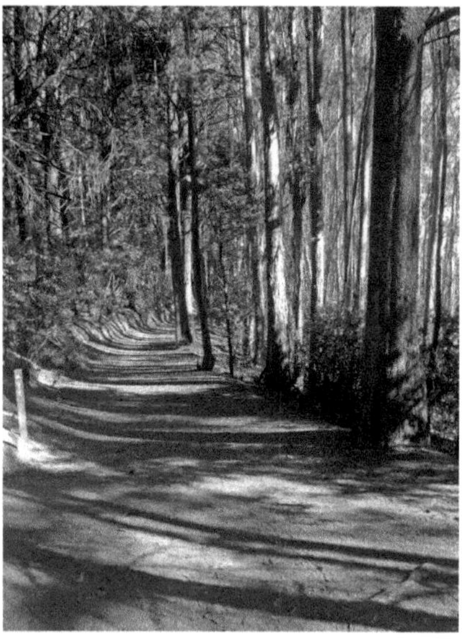

Left: An orderly row of nonnative Monterey cypress trees lines the West Ridge Trail in this photograph taken around 1940. *Courtesy East Bay Regional Park District archives.*

Right: The same colonnade of cypress trees is seen in this 2015 photo. Did Frank Havens and the Realty Syndicate plant them to welcome real estate buyers to Redwood Peak? *Amelia Sue Marshall photo.*

Mule Team Borax fame, Havens operated the Realty Syndicate, selling lots throughout the hills. His other prominent ventures included construction of the Claremont Hotel and the Key System of light rail transit.

To make the Oakland hills more attractive to prospective real estate buyers, Havens's Mahogany Eucalyptus and Land Company planted nearly 3 million Tasmanian blue gum eucalyptus, Monterey pine and Monterey cypress seedlings on three thousand acres. When the fast-growing eucalyptus proved to be unsuitable for lumber, Havens shut down his nurseries and sawmills. But the ecosystem of the hills was left greatly changed.

Havens was not the only person who planted trees. Beginning around 1906, Progressives such as President Theodore Roosevelt and Gifford Pinchot, head of the United States Forest Service, began a program of public education around conservation of trees. Arbor Day tree planting was undertaken worldwide.

The poet Joaquin Miller was a tree-planting enthusiast as well. Not only did he add pines to the landscape, but he also brought acacia trees, with their yellow flowers cascading pollen. With volunteer (or conscripted) labor from his endless parade of visitors, Joaquin Miller is credited with having planted seventy-five thousand trees in the Oakland hills between 1886 and 1913.

Initially, the poet planted fruit trees of all types—olive, apple, pear, plum and cherry—and also citrus and banana. The latter did not thrive.

Settlers of the Middle Redwoods: Redwood Canyon

Following the clear-cutting of the ancient redwoods, remnant logging towns remained in the deep canyons east of Oakland. Along the old Tupper Road, Redwood Creek tumbled eastward downhill to San Leandro Creek. Ranch houses were situated in the cool and scenic valleys. Cattle grazed on the sunbaked hillsides above. The dairy industry provided a means of livelihood for a gradually increasing population of settlers in Redwood Canyon in the years between 1860 and the advent of the automobile. Ranchers could carry milk by the wagon load along the Tupper Road, over the hill to Brooklyn, San Antonio and Oakland.

Little is known about the original farmers who occupied the valley that is now the Stream Trail area of Redwood Regional Park. Here the Middle Redwoods had stood. Second-growth trees had begun to regenerate when the O'Brien family first farmed the valley. Fields were clear for planting where the sawmills had stood along the stream

The McNee Ranch

Alexander McNee (1878–1938) was the son of Duncan McNee, the "early California land baron"[145] who owned a vast ranch—800,000 acres—along the San Mateo County coast, including Montara and Half Moon Bay. Alexander was the largest landowner in the area at the inception of the East Bay Regional Park District. He also had investments in acreage in the Moraga redwoods. Alexander lived in the East Bay until his death, in Berkeley, in 1938.

The Peterson and Alves Homesteads

These homesteads were identified on an 1878 map and is cited in the 1982 park district cultural resources survey of Anthony Chabot Regional Park.[146] It was located at today's MacDonald Trail staging area, on the west side of the confluence of Redwood Creek and an unnamed tributary. Banks, who wrote the survey document, noted that a Thompson and West map shows the Peterson homestead on the east side of this confluence, with the neighboring homestead called Alves. The homestead also appears on an 1899 map.

The Redwood School

The rough society of Redwood Canyon during the logging years apparently did not see the need for a school. Nearby, in Moraga, when the Elam Brown party first arrived in 1847, it found several families already in residence. The Willow Springs School was established to serve local children. The no-nonsense "Parson" Brown used this school for his Sunday sermons.

Joseph Lamson described how Mr. Sage, the Willow Springs schoolmaster, served as an armed escort for Parson Brown's wife and children, plus fifteen others, when they rode up to Knobcone Pine Ridge, above Indian Valley, to gather huckleberries. "As this mountain was known to be frequented by grizzlies, Mrs. Brown was rather desirous of having a guard in case of attack."[147]

One early school was founded in about 1880 in the now-vanished town of Redwood, according to a 1964 article by Dodie Livingston in the *Oakland Tribune*:[148] "Redwood [School] was founded 104 years ago in a beautiful

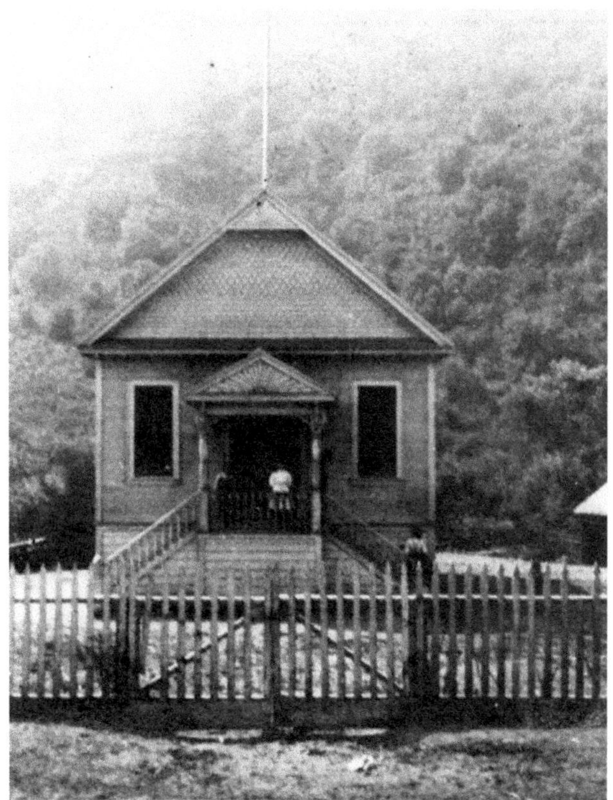

Right: The Redwood School is believed to have been reconstructed at its present location after construction of lower San Leandro Reservoir flooded the original town of Redwood. *Courtesy East Bay Regional Park District archives.*

Below: Henry Hauschildt, schoolmaster at Redwood School, stands near today's Stream Trail entrance to Redwood Regional Park. *Courtesy East Bay Regional Park District archives.*

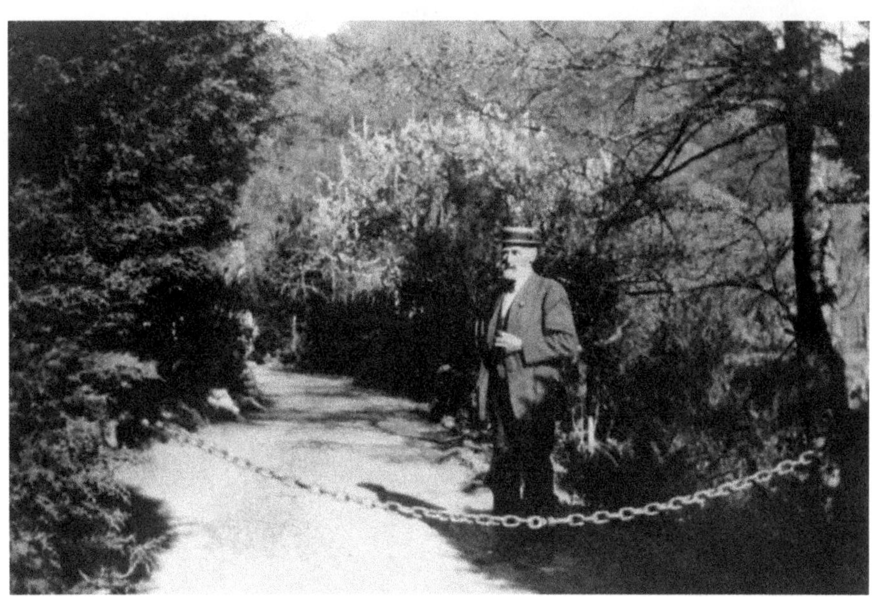

valley which just at the turn of the century became San Leandro Reservoir. The district resettled in the present school house."

Undated photographs of the newly constructed Redwood School building and its schoolmaster, Henry Hauschildt, suggest that this was the original one-room schoolhouse near today's Redwood Regional Park entrance. By 1939, the schoolhouse had been moved into a new building on the hillside. A fire station was constructed in the 1930s between the park entrance and the school.

Between 1940 and 1960, when there were still many private homes along Redwood Road between the Oakland city limits and Pinehurst Road, the one-room school continued to thrive.

A 1956 *Oakland Tribune* article by Bill Stroebel[149] describes three Alameda County single-room schools then still in existence. Stroebel noted that Redwood School teacher Ruth Bellini shared the belief that students got more individual attention and actually learned more there than in a larger school:

> *Mrs. Bellini, who is likely to be found sprawled out on the floor with her charges, working on a mural, is assured of an apple for the teacher. There is an apple orchard on the property.*[150]
>
> *In addition, the youngsters have a cherry tree and their own garden—which helps considerably, because she usually makes lunch for them and serves lemonade in the afternoon. Last year a deer wandered into the schoolyard, and until a wire fence was installed last year, the children could collect trout for pets from a nearby stream.*

This idyllic setting for children was too good to last. In 1964, the Castro Valley School District shut it down, citing declining enrollment. "Redwood Canyon is owned, for the most part, by the East Bay Regional Park District and the East Bay Municipal Utility District. Gradually, over the years, efforts to turn the property back into watershed have forced out homeowners."[151]

The four students in Mrs. Bellini's final class in 1964 were boys. Three brothers were from the Morrissey family—Keith, age seven; Kirt, eight; and Michael, nine—as well as John Littlejohn, nine. They were sent to larger schools in Oakland, where there would be no apple trees, deer or trout.

The hillside school building remains and is used by the park district for offices.

John (Marcillio Borge dos) Reis and Lydia Ramos Reis

Marcillio (who preferred to be called John) and Lydia Reis operated a ranch and dairy in Redwood Regional Park between 1908 and 1960. After John's death in 1966, Lydia remained in the area. Their house at the entrance to Redwood Regional Park still stands[152] and is now used as a staff residence.

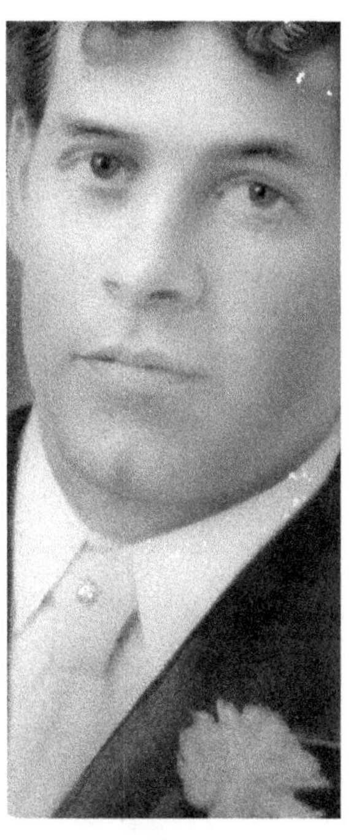

Marcillio ("John") Reis was a presence in Redwood Canyon from 1907 through 1966. He planted fruit trees, built stone fireplaces and stairways and operated a dairy at the site of Piedmont Stables. *Courtesy John Whatley.*

One of their three children, Lydia Reis Whatley, continued to live in the area until she died in 2005 at the age of 101. Lydia Whatley, who spent her career working for the Oakland Unified School District, provided handwritten notes for her family, detailing their remarkable history.

John Reis was one of the many Portuguese immigrants who came to the East Bay to make a home. Arriving in California from the Azores island of Sao Jorge in 1902, Reis joined his elder brother, Paul, who worked at the gunpowder plant in Hercules.

John took up dairy farming, a profitable line of work, when he was hired by Anderson Dairy, located at the foot of Redwood Road, near today's Lincoln Square shopping center. As a child on Sao Jorge, he had taken care of livestock from the time he was eight years old. He had begun learning to cook when he was even younger. Later in life, he would be renowned as a gourmet chef who cooked for church and charitable events. "Anderson Dairy was a plum job—Sundays off," Lydia Reis Whatley wrote. "Maggie the cook fixed chicken for Sunday dinner—but the rest of the week they had fish, cabbage and potatoes—but a good breakfast of beans, stew and hotcakes....Next to Anderson's Dairy was the Alma Mine [operated by] Stauffer Chemical. They mined it for some time after he was married in 1908. He was able to buy a team of horses and a wagon, so he left the dairy and worked hauling ore [from the mine] to the chemical plant."[153]

John Reis had met Mrs. Ramos, who owned the Melrose Creamery. She operated a dairy in Redwood Canyon at the site of today's Piedmont

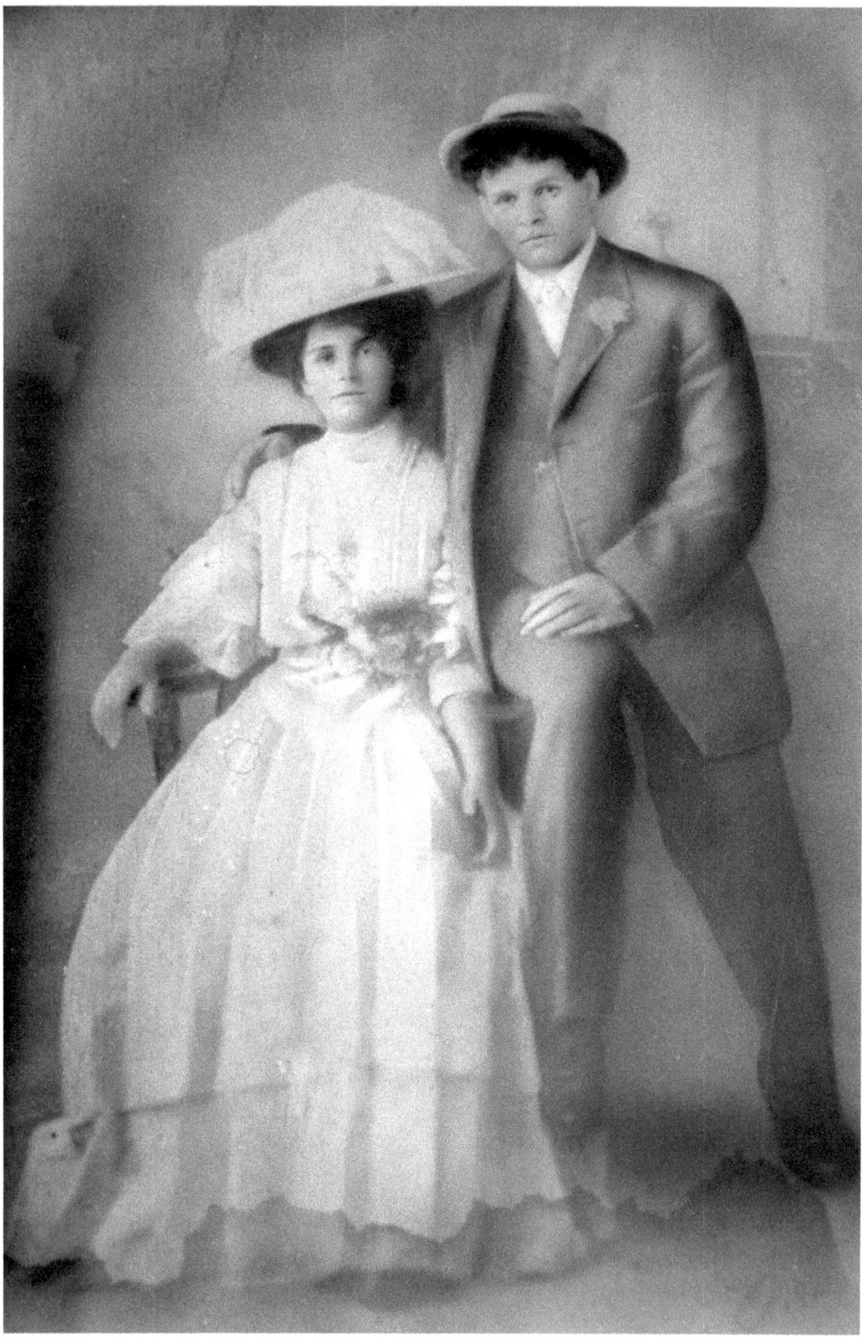

Marcillio ("John") Reis weds Lydia Ramos in 1908. They planted the orchard and operated a dairy in Redwood Canyon; their family members have lived in the area for three generations. *Courtesy John Whatley.*

Stables. One day, John rode up to help them with their cattle. There he first encountered Lydia Constance Ramos, the dairy owner's daughter. "My father saw her in the pig pen, holding a baby pig," Lydia wrote. "He said she was the prettiest thing he ever saw. They were married a year later, in May 1908."

The newlyweds lived in nearby Allendale, where they had three children—George, Mabel and Lydia. John worked in the Leona Heights mines and used his team for plowing and grading. Lydia Whatley recalled that "all of East Fourteenth [Street], and also Hopkins Street,[154] had teams excavating, putting in new streets, so he was making out very well....In 1922, he was able to lease a ranch in Redwood Canyon [now Redwood Regional Park] from East Bay Water Company and Havenscourt [Frank and Lila Havens], who owned the upper ranch."

John Reis leased 1,800 acres. Some of this land had been the property of the Bridges family prior to 1909, before it had been acquired by Frank Havens and the water company. Adjacent parcels were leased by the Piedmont Trails Club.

Lydia Whatley recalled that her father kept thirty milk cows on his ranch. He sold milk to Golden State Creamery. He raised enough hay for his own stock, plus a surplus that was sold to other dairy farmers.

In addition to being a dairyman, rancher and chef, John was a stonemason. He constructed fireplaces of stonemasonry at the Redwood Roundup Saloon and other nearby buildings.[155] The stone steps to the old Redwood Hunting Lodge, a short distance from John and Lydia's homestead, were likely the work of John Reis. The lodge itself was located on the south side of the prefabricated building that serves as the Redwood Regional Park headquarters.

The massive stone fireplace at the Big Bear Tavern, which stood across Redwood Road from the park entrance, was likely constructed by John Reis as well. The park district demolished both tavern buildings, the Big Bear in 1965 and the Redwood Roundup, in 1990.

Lydia Whatley recalled that while the family lived in the flatlands, in the close-knit Portuguese community, the ranch in Redwood Canyon was an important part of her childhood in the years following World War I:

> *We lived in Allendale on the corner of 38th Avenue and Dale* [Place].... *Dad belonged to the Allendale Business Association....Their purpose was to help to improve the neighborhood, especially the poor. They celebrated all the holidays with dinners* [to raise] *money for eyeglasses and health care*

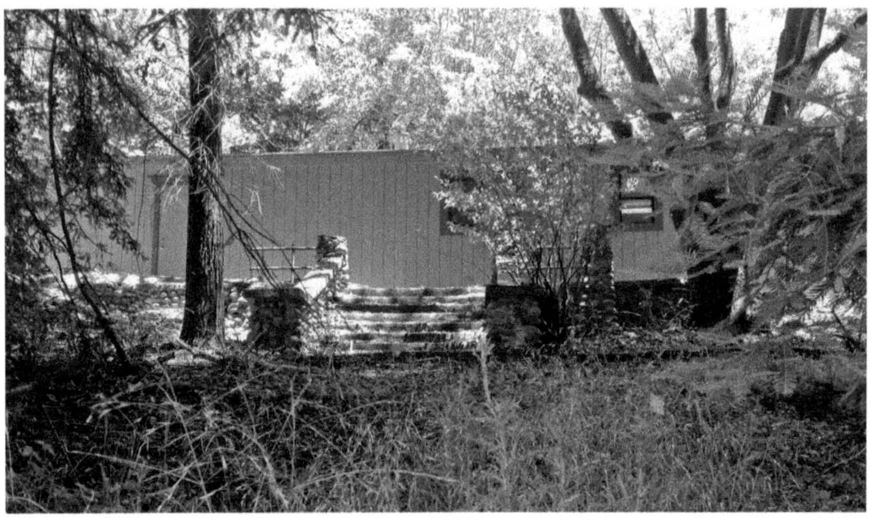

The stone steps are all that remain of the Redwood Hunting Lodge, demolished after the 1989 Loma Prieta earthquake. It is likely that John Reis was the stonemason who built them. *Amelia Sue Marshall photo.*

> *for the children in need. Dad did all the cooking—turkey, beef, and chicken. He donated the meat from the ranch.*
>
> [Across 38th Avenue from our home] *was a large vacant lot. Every Christmas, Dad would cut a large Redwood Christmas tree and put it in the lot. The* [men from the Allendale Business Association] *would decorate it. Just before Christmas, they would have a special day, with Santa Claus giving candy canes, apples, oranges, and bananas to the children. It was fun for us, and lots of excitement.*
>
> *Also a few weeks before Christmas, Dad would invite a group of men from the* [World War I] *veterans' home to the ranch to help him cut Christmas greens from the Redwood trees. The fellows enjoyed getting away from the home, the good food, and the spirit of Christmas. They cut a big hay wagon full of greens, and Dad took them to Podesta and Baldocchi*[156] *so they could make wreaths and garlands to decorate the streets. This took about a week. He and grandma cooked big meals for the men. They really enjoyed it, and looked forward to coming back each year.*
>
> *It was fun for us kids riding in the hay wagon and helping bundle the greens. Lots of fun jumping into a big pile of greens!!*

John Reis served as a volunteer sheriff's deputy in Redwood Canyon, starting about 1919. From 1935 through 1939, he worked as a labor foreman on the Bay Bridge construction project. "He was [on the Bay Bridge] from start to finish," Lydia Whatley wrote. "He was the last man there, until the work was completed. He rode in the first group of cars to cross the bridge." Reis served as the volunteer fire chief in Redwood Canyon from about 1940 to 1959. During World War II, he was the Redwood Canyon civil defense warden from 1941 to 1945. In a 2016 interview with the author, grandson John Whatley, a longtime Oakland resident, recalled visiting his grandfather when he was a child:

> *My grandfather had a lot of things going on in Redwood Canyon. He kept cattle, planted the orchard, and maintained water tanks. He did marvelous stone work. He served as a volunteer fireman at the firehouse that was close to the one room schoolhouse. He had the apple orchard near the firehouse. When I was a child in the 1940s, we used to pick the apples.*
>
> *My grandfather treated his animals well. The butterfat content of the milk from his cows was so high that the Golden State Creamery offered to set him up in a different location after he got out of the dairy business in Redwood Canyon.*
>
> *Grandfather John Reis was a friend to all the residents of Redwood Canyon.*

Marcillio John Reis is buried at the Mountain View Cemetery in Oakland.

The Montiero Ranch

While little is known about the Montiero family, who ran cattle in the hills above Redwood Canyon, their name lives on through the Montiero Trail. Their ranch house stood a few hundred feet north of Redwood Road, on a side road that is used as access to the Big Spring. Emerging from the earth above the creek channel, the spring provides year-round water for Piedmont Stables and several private homes.

The names Walter, George and Edward Montiero are listed on the 1949 documents from the time their property was sold to the park district. General Manager Richard Walpole intended to use their house, valued at $5,000, for a ranger station. It was used for park offices until it was demolished at the same time as the Redwood Inn (Redwood Roundup), in 1995.

Roadhouses and a Bordello

During the early years of the twentieth century, Redwood Canyon was a rural enclave with affordable homesites for cowboys who worked on the surrounding ranches, miners and other laborers. Lots along Redwood Road were kept in private ownership, while the East Bay Water Company kept title to the watershed along San Leandro Creek.

There is little documentation about what went on in Redwood Canyon during the Prohibition era. Clearly, the residents did not wish to rid their community of its moonshiners and saloons.

Today, with few private homes remaining in Redwood Canyon, it is hard to imagine the time when it was a bustling community. The Redwood Hotel stood on the west side of Redwood Road, across from the park entrance. A retaining wall of large stones, distinctly different from the surrounding roadside, marks the place where it once stood. According to local lore, it was popular with hunters and fishermen.

Within the main gate of Redwood Regional Park, just beyond today's park office, stood the Redwood Hunting Lodge. Before the park district acquired the land, a farmer named Mr. Fowler was the caretaker for the old lodge. Fowler had a mule named Mary Ann.[157]

According to former park supervisor Dee Rosario, the building had ornate Belgian tiles on the floor. It was irreparably damaged by the Loma Prieta Earthquake in 1989 and was demolished the following year.

Pure Spring Water

In addition to pure water from Redwood Creek, the surrounding hillsides featured a number of springs that residents tapped for domestic use. The underlying rock in the area includes porous serpentine, through which springs percolate.

The January 1931 Oakland city directory lists the Big Bear Spring Water Company ("Wilkinson's, distr.") on Redwood Road, phone number ANdover 4146.

According to local horseman Larry Brickell, the Alhambra water company at some point purchased water rights to the spring and built a large tank on the north-facing canyon wall. Guides from the Big Bear Stables would lead riders up a short loop trail around the tank.

Eventually, the two agencies—the park district and East Bay Municipal Utility District (EBMUD)—resolved the water rights question so that the Alhambra company could no longer tap the spring. "I remember the Alhambra water spring issue because people came to the pipe from the spring to get free water for years," Jerry Kent recalled.

The 1943 Oakland city directory no longer listed any private water companies on Redwood Road.

The Big Bear Tavern: A Jazz Landmark, Vanished

The Big Bear Tavern[158] may have started as a speakeasy during the Prohibition era (1920–33), but perhaps the tavern was much older. With a good source of spring water close by, it could have gone into business any time after the post–Gold Rush logging era. Although it attracted a rowdy clientele throughout its history, the place thrived until the park district bought the property and demolished the tavern in 1965.

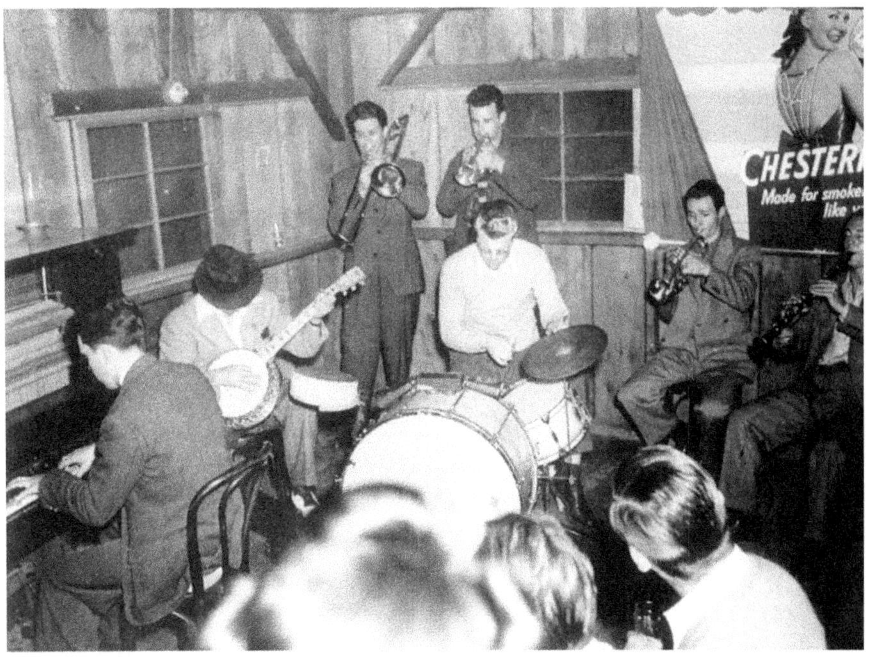

The Big Bear Tavern was the epicenter of the traditional jazz revival of the 1940s. Here Lu Watters, Turk Murphy and the Yerba Buena Jazz Band jam after hours. *Courtesy Dave Radlauer.*

The first newspaper accounts of bad behavior there came during the 1930s. Authorities from two counties raided the Big Bear in 1935 to confiscate illegal gambling machines.[159] The Big Bear owner was identified as one "E. Smith" in news reports. Alameda County district attorney Earl Warren was at that time engaged in a program to rid his entire jurisdiction of vice.

The Big Bear changed hands. It was sold by L.J. Bischoff in July 1936.[160] In 1938, one Jack Remers was shot in the arm during a dispute at the Big Bear. Remers and some friends had brought in their own beer and had been told that they could not drink it on the premises.[161]

However, the Big Bear Tavern remains best known for its role as an incubator of a new style of jazz played by multiracial bands in the 1940s. In a rebellion against the bland big-band music of the World War II era, local musicians Lu Watters and Melvin E. "Turk" Murphy invited national celebrities, such as New Orleans trumpet player Bunk Johnson, to jam in this rustic setting. "Mr. Murphy was the trombonist in the Lu Watters Yerba Buena Jazz Band, which was formed in 1940 in after-hours sessions at the Big Bear Tavern in the Oakland hills," the *New York Times* reported.[162]

According to Bay Area jazz historian Dave Radlauer, "The Traditional Jazz revival began at Big Bear Tavern in the Oakland Hills. The tavern was an old fashioned roadhouse containing stables for horses and a dance floor for 400 people. Jam sessions at Big Bear draw jazz musicians from all over the Bay Area to what clarinetist Bob Helm called, 'an experimental laboratory of jazz.'"[163] The tavern was immortalized by Lu Watters and the Yerba Buena Jazz Band in a piece called "The Big Bear Stomp."[164]

Adventurous students from nearby Mills College drove over the hill and down the canyon on dates. Perhaps the first recorded auto burglary in Redwood Regional Park occurred when the baggage of Mills student Sally E. Carnew, containing $600 worth of jewelry, was stolen from an unlocked car in the Big Bear parking area.[165]

The Big Bear Tavern stood on the west side of Redwood Road, between the Redwood Hotel site and Piedmont Stables. In addition to the large stone fireplace, there was a café, a bar and the large dance floor. Directly across the road was the Big Bear Stables, operated by the Bradley family until the early 1970s, when the park district seized that property by eminent domain. The park district demolished the tavern itself during the 1960s.

"The Big Bear Tavern conflict with the park district centered around the fact that their septic tank was leaching into Redwood Creek and San Leandro Reservoir," recalled Jerry Kent. "I was told that General Manager William Penn Mott and EBMUD general manager John McFarland had reached an agreement to remove private septic 'uses' in the canyon. The park district was buying and demolishing homes and other structures on one side of the road, and EBMUD was buying the other side. The bar and stables did not fit into Mott's vision of Redwood Creek or the heavily used Stream Trail area of Redwood Park."

After the tavern was long gone, in 1999, a mock service station set was built in the Big Bear staging area for the film *True Crime*, directed by Clint Eastwood. Many a visitor to the Redwood Regional Park administrative building nearby has been puzzled by the "Redwood Garage" sign that hangs above the workshop. There was never an actual service station in the vicinity; the sign is a remnant of the movie set.

Ruby's Bordello

To round out the recreational amenities of Redwood Canyon, Ruby's bordello stood in a discreet location, upslope from Redwood Road on the west side, across the road,[166] just south of the turnoff to Pinehurst Road. Other than a 2007 *Montclarion* column, "Looking Back," by Erika Mailman,[167] no written documentation has survived about the brothel or its proprietor.

Lydia Reis Whatley, however, described to Dee Rosario[168] the role that the bordello played in the community. "Guys who worked in the Stauffer Chemical mines would stop at Ruby's on their way to and from work, Lydia told me," said Rosario.

"I used to tease my wife, that if she didn't treat me well, I could always go down to that house of ill repute," recalled Lloyd Graham, longtime operator of Piedmont Stables.

Undertaking a search for the ruins of the bordello around 1998, Rosario clambered up an overgrown slope that had probably once been a driveway. There he discovered the foundation wall and steps that once led to the basement. The forested hillside will continue to keep the secrets of Ruby's bordello.

The Redwood Roundup

Conveniently located next door to the Montiero brothers' ranch house was the establishment that was called, at various points in its history, the Roundup, the Redwood Roundup or the Redwood Inn. Oakland city directories from

1940 and onward list the Roundup as well as the Big Bear Tavern under "Night Clubs."

The original Roundup was owned by George and Helen Willis in the 1940s and '50s. After they retired, their son, Don Willis, and his wife, Solange, took it over and renamed it the Redwood Inn. "In the late '50s—1958, '59, '60—that was a rocking place," recalled Bob Schultz, an artist who had been hired by William Penn Mott to design exhibits at Oakland Childen's Fairyland. "My uncle Phil worked as a bouncer at the Roundup. I liked to go there to hear Kurt Okie and his Oklahoma Cottonpickers. It was quite the country and western music venue."

Up until the 1970s, it was not unusual to see horses tied to the railings outside while their riders got a meal and perhaps a drink or two. The atmosphere changed in the 1970s. A mirrored "disco ball" was added to the décor. "The honkey-tonk era ended," Schultz noted. "The new owner moved in an organ. He would play country music with a Lawrence Welk beat, dressed in his powder blue cowboy hat and blue silk scarf."

"I was there many times, circa 1970s," commented Oakland resident Art Neely. "They had a good steak there. The new owners tried to make it into a dance club. The old place was quaint and cozy. The dance club did not last."

By the 1980s, the East Bay Regional Park District was using the old Montiero Ranch buildings for staff offices. Forester John Nicoles knew the area well. "The Redwood Inn was actually a *great* place for a shoe-leather steak and lots of atmosphere," Nicoles recalled.

Regardless of the style of entertainment or cuisine, water rights proved to be the more urgent issue. "All the buildings around the Redwood Roundup were served with spring water that came off park district property," Nicoles explained.

> *The owner of the Redwood Inn complained that the district was threatening to cut off the water. There was a historic verbal agreement but nothing in writing.*
>
> *What finally did in the Inn, however, was the contention that their restrooms were leaking sewage into Redwood Creek. I cannot confirm or disprove that the leakage was occurring, although the location of the restrooms certainly made it possible. I think the matter might have been easily repaired, but the district did not view the Inn* [in a] *favorable light and wanted it gone. Whether the sewage issue was real or not, it was used as a lever, and the Inn was closed.*

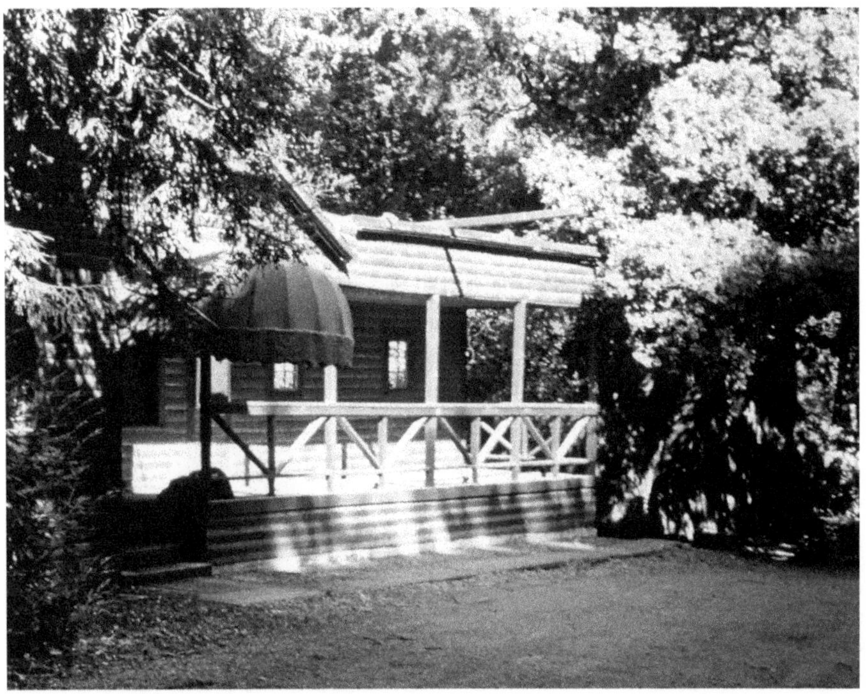

The Redwood Roundup was originally a tavern serving the Montiero Ranch cowboys. During the 1960s and 1970s, it became a hot spot for western music. *Courtesy East Bay Regional Park District archives.*

Prior to the demolition of the Redwood Roundup in 1995, the park district engaged a consultant to assess the cultural value of the building. It was not deemed worthy of saving.[169]

The foundations of the old Monteiro ranch buildings can still be found, near where the Montiero Trail ascends the hill from the site of the Redwood Roundup.

SETTLERS OF THE MORAGA REDWOODS

The Town of Canyon

Canyon is an isolated village in the Moraga Redwoods, just one mile from the Oakland city limits. Its tapestry of historic events, folklore and odd circumstances has been the subject of two books, a master's thesis and innumerable articles.[170] We will tell the broad outlines of its story here.

In the deep, fragrant canyon, Hiram Thorn and William Hamilton built their sawmill in 1852. While the precise location of the mill is not known, longtime Canyon School teacher Esperanza Pratt Surls believes that the Canyon school site is the likely location. "It stands to reason that the mill would be here," Surls explained. "It is a large, flat area along the creek."

Hiram Thorn's toll road extended up Pinehurst Road, through today's Huckleberry Botanic Preserve and over the ridge crest. Wagons drawn by horses or oxen carried loads of rough-cut lumber westward down today's Thornhill Drive toward the Sausal Creek Canyon on the way to the *embarcadero* at the foot of Fourteenth Avenue.

By the time the Thorn mill closed around 1860, a rough frontier town had grown up around the mill. This was as close as civilization came to the pioneer McCosker, Pereira, Manes, DeAvila, Bordersen and Hansen families, as well as others.

Unpaid Native men who worked for Joaquín Moraga built the first house in Canyon in 1830. It was acquired in 1889 by Killian Messmer, who was known as "Charcoal John." According to Albert McCosker, Messmer made charcoal and sold it in Oakland. He left Canyon under a cloud, accused of being a "bad neighbor." This vague description alluded to trouble with the law due to horse stealing and mortgage fraud.[171]

With the redwood trees gone, the pioneers made an effort at farming, particularly on the upper slopes, where the grass-covered hilltops were free of stumps and logging debris. The steep slopes, however, were not amenable to tilling. And the soil did not support crops. Running dairy cattle was seen as the most profitable use of the land.[172]

Robert Manes built the second house in Canyon, at some point after the lumbering era. After his death in 1889, his widow, Marianne, and her sister, Susan McDonald, engaged in legal disputes for six years over the ownership of 160 acres. The property was eventually acquired by EBMUD.

The Saloon Era

In the 1850s, there were five saloons, one or two brothels and a dance floor, reported Canyon native Annabelle Williams in her book *Handmade Lives*. Local lore—and, in some cases, public records—recounts land battles, highway robberies, lynchings, beer-drinking horses and pet coyotes.[173]

A stagecoach served Canyon during the latter decades of the nineteenth century. One of its drivers, Henry S. Jones, had a house on the Thorn Toll

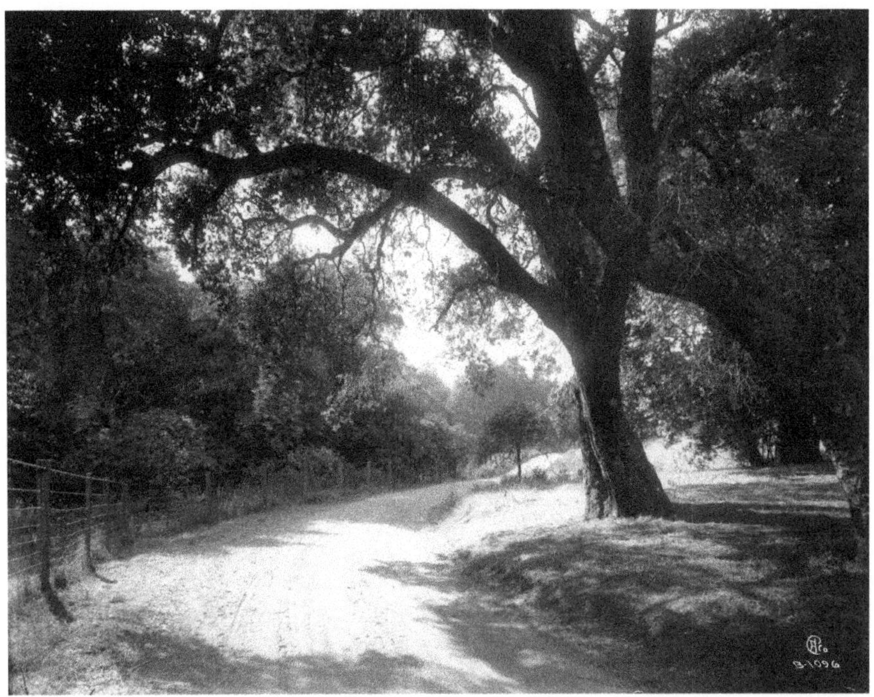

The Hangman's Tree in Canyon was the focus of many folktales until it died and collapsed around 2000. *Courtesy Moraga Historical Society.*

Road near the stage stop at Andrew Hansen's Hotel.[174] The driver was known locally as "Whispering Jones" because of his booming voice.

Hikers, hunters and campers rode the stagecoach to visit the area. They lodged at Hansen's Hotel or at Happy Camp, run by a man named Smith. According to longtime resident Gladys Shally, Hansen's was actually a saloon, with upstairs rooms for rent.[175]

County records show a long series of legal battles over Hansen's homestead, beginning in 1890 when he filed for divorce against his wife, Mary. Hansen agreed to sell his land to real estate investor William J. Dingee. Eventually, in 1903, the courts declared Mary Hansen and a W.M. Watson (likely a creditor of Dingee's) the rightful owners. Through a rapid series of subsequent sales, the land became the property of the Realty Syndicate. It would eventually become one of the railroad subdivisions.[176]

Henry ("Harry") Bird operated another saloon and hotel. He was in 1886 appointed the first postmaster of Moraga. Bird's Hotel was located close to the present Canyon School.[177]

According to Gladys Shally, the British-born Bird was well liked in Canyon. However, he was involved in a protracted feud with James and George Wallace. These brothers operated Camp Wallace, at the site of today's Canyon School playground, across the road from Bird's Hotel. Bird believed that the brothers were trespassing on his property. They were also evidently in competition with his saloon. In 1887, George Wallace built a fence around his establishment. Bird tore the fence down. He then shot Wallace in the back near the hotel in broad daylight, with witnesses.[178] After he was remanded to San Quentin, Bird deeded his property in Canyon to his attorney, Eli R. Chase.[179]

The most notorious liquor dealer in Canyon was Isaac Bottomley. He augmented his retail liquor sales operation in Oakland with a saloon in Canyon and extensive real estate dealings there.

Margaret Bottomley, wife of Isaac, in 1901 purchased lands that had been the pioneer DeAvila family farm. It was there that Isaac Bottomley ran a saloon that was noted for its rough clientele. According to McEwan, "Although Canyon was not an overly genteel community, cock-fighting and the accompanying gambling had not been a regular feature of the Canyon milieu [before Isaac Bottomley established his saloon]. Canyon was becoming a weekend gambling spot."[180]

The Bohan family, whose patriarch, John, was a blacksmith in Oakland, became involved with the Bottomley family through real estate deals and with the McCosker family by marriage. John Bohan had acquired property at Acalanes Ranch. Upon his death, his widow, Maria, and son Michael began investing in Canyon real estate. In 1906, Michael purchased ten acres at the Pinehurst end of the Bottomleys' land. Included were a saloon, two dwellings and a barn.[181] Later, Patrick Bohan, brother of Michael, married Catherine McCosker, daughter of the pioneer couple Patrick and Catherine McCosker.

Isaac Bottomley continued his profitable saloon business. He built the original Redwood Inn up the canyon, about two hundred yards below Hansen's Hotel and Saloon.[182]

The Railroad Years

The years when the Sacramento Northern Railway was in operation saw farmland subdivided into residential lots. By the time the railroad ran through Canyon, fifty years had elapsed since Thorn's Mill was closed.

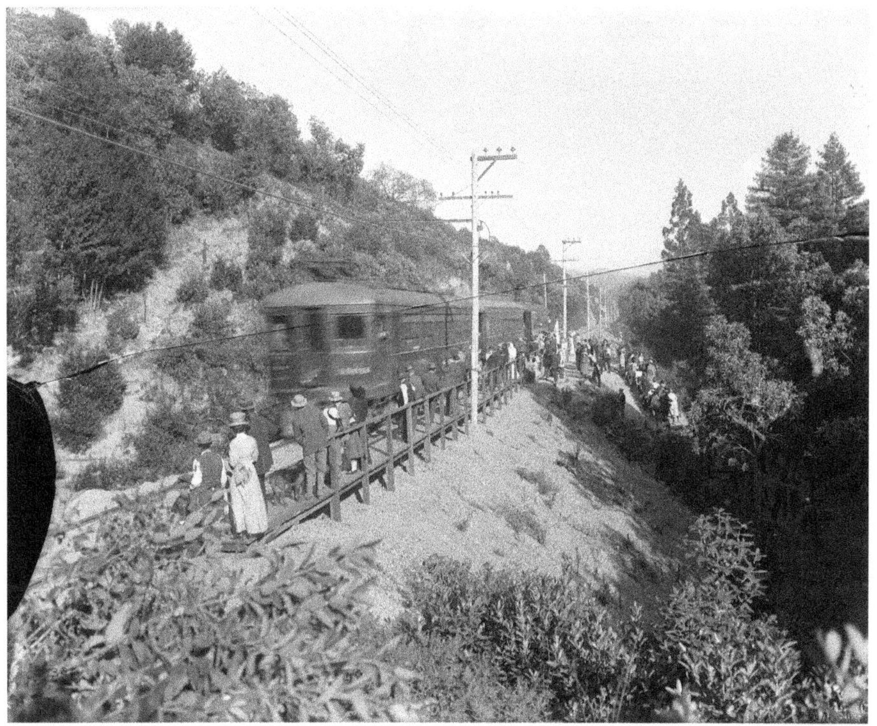

Passengers line up to board the Sacramento Northern near the town of Canyon. *Courtesy Western Railroad Museum.*

The second-growth redwood trees shaded a scenic area suitable for vacation homes.

From 1914 to 1929, residents built a school, a store and an inn and started a community club. Electric and telephone service allowed families to take up year-round residence.[183]

The Redwood Inn[184]—with its store, post office and dance hall—was Canyon's social center. Town meetings were held in the dance hall, which had been built by George Lydiksen for George Williamson, the proprietor.

The Canyon Community Club was organized in 1924 as the town's governing body. It had four officers and held open meetings.

The post office was opened on August 23, 1922, but the town was initially named Sequoya. It was renamed Canyon in 1927 to avoid confusion with another California town of the same name.

Protestant residents began holding services at the Inn, while Catholics rode the train to Mass at St. Leo's Church in Oakland.[185]

With the coming of the 1930s, the Depression era brought changes to Canyon in the form of increased population. Some San Francisco and East Bay families lost their homes to foreclosure and had to move into their summer homes. The population of Canyon rose to about four hundred.[186]

Under those circumstances, the Sacramento Northern Railway provided a lifeline. Still, during hard times, those without an automobile—wives, in particular—found themselves isolated.

Post–World War II Housing Shortage

Surrounded as it was with protected watershed lands owned by EBMUD, Canyon maintained its secluded and insular atmosphere. Urban sprawl was impossible. Increased population density was a different story, however.

During the World War II years, the population of the Bay Area increased greatly as working people sought jobs in the shipyards and wartime industries. Naturally, most stayed after peace was restored. The baby boom began. Soon there was a housing shortage.

In *Its Name Was MUD*, the official history of the East Bay Municipal Utility District (EBMUD) published in 1970, author John Wesley Noble explained how utility managers viewed the population of Canyon:

> *Canyon lies astride the creek that flows directly into Upper San Leandro Reservoir.…The colony of summer homes that resulted* [from the railroad subdivisions] *depended on septic tanks to handle sewage and was not intended for year-round use. However, when housing of any sort was at a premium…*[during and after] *World War II, year around residents took over the summer houses and began to create a definite hazard to the downstream water supply. Whenever winter rain soaked the ground, the old septic tanks overflowed directly into the creek.*[187]

In 1951, EBMUD adopted a sanitation policy that, in part, called for the acquisition of any private property on watershed lands. This resulted in the destruction of the town of Valle Vista and the purchase of any Canyon property that came on the market. "Surprisingly, the utilities district initiated its land-purchase program without first taking a bacteriological test of the contamination levels of the water in the creek," noted John Van Der Zee.[188]

Van Der Zee quoted Gordon Laverty, an EBMUD sanitation engineer who later became the water services manager: "We just didn't sample and plot things in 1950–55."

Canyon residents were concerned about the depletion of their close-knit community. There was a precipitous drop in enrollment in the Canyon School. The number of students dropped below the state-required minimum for the first time since 1918.[189] "[W]hen people from Canyon, passing by the vacant, grassy hollow on the highway to Moraga, look at what remains of Valle Vista, they see the future East Bay MUD has in mind for them," wrote Van Der Zee.

After the utility managers attempted to buy the Canyon Store, angry residents formed the Canyon Trust and raised money to buy the store. An atmosphere of animosity prevailed.

EBMUD versus the Hippies of Canyon

By 1967, the hippie counterculture was in full bloom. The rustic charm and unregimented atmosphere of Canyon attracted long-haired youth seeking to go "back to the land" without having to travel far.

Canyon native Annabelle Williams described the ethos of the times: "Many children were born. Natural home births became occasions for ceremony. The community parented all the children communally, with great attention paid to the operation of the Canyon School."[190] Storied rock music groups, creators of the "San Francisco sound," were often found enjoying the groove of rural Canyon. Van Der Zee described the newer Canyon residents:

> *In the early nineteen sixties, Berkeley became Canyon's window on the outside world. Half of Berkeley, it was said, wanted to move to Canyon....They were graduate students and university staff, lab technicians and teaching assistants, street people, kids who had...now returned....*[U]*nlike the cult-and-commune faddists...most of* [the Canyon hippies] *remained independent of each other by continuing in school or maintaining jobs outside the community.*
>
> *Those who could afford to bought land and old houses, usually with the last cent of their savings, and these homes became homes to their friends, too....If one of these friends decided to stay in Canyon and wanted to build a shelter or a one-room shack, he could usually find someone who would show him how to build, help him find materials, and lend him tools.*

> *This helped him produce some remarkably inventive and individualistic architecture: geodesic domes, tree houses, and lightweight, cheap, or castoff materials used in original ways.*[191]

The Canyon Community Club had functioned as an informal means of local governance since the era of the railroad subdivisions. Now some of the newer civic-minded residents began to participate. Both older "straight" residents and newer, hipper ones engaged in a series of community meetings with EBMUD managers. The community largely viewed the utility as intransigent. "In spite of its 'tread lightly' approach, however, there was no doubt that EBMUD's ultimate goal was to obtain complete title to the Canyon watershed area," wrote Noble, expressing the point of view of the utility managers.

The Dawn of Community Ownership

The Canyon Store and Dance Hall had become dilapidated. Its owner was Mrs. Holmes, an elderly woman who had been unable to operate or lease the property. According to Van Der Zee, she had reluctantly agreed to accept the EBMUD bid for $55,000. She assumed that no one in Canyon could offer her more.

> [W]*ithout open discussion and under a veiled wording of its agenda, the East Bay MUD board of directors voted to acquire and tear down the Canyon General Store, post office, and dance hall, cutting off access to Pinehurst Road and eliminating the heart of the community's social life.... The residents were shocked and angry. Though the dance hall was no longer operating, the post office and store were still the one place where everyone went daily, where news and gossip were exchanged and notes posted.*
>
> *That evening, three of the older residents who had remained active in the community called an emergency meeting at the Canyon School. They agreed to form a Store Trust, pledging* [their own savings].[192]

The deal between Mrs. Holmes and the utility had not been finalized.

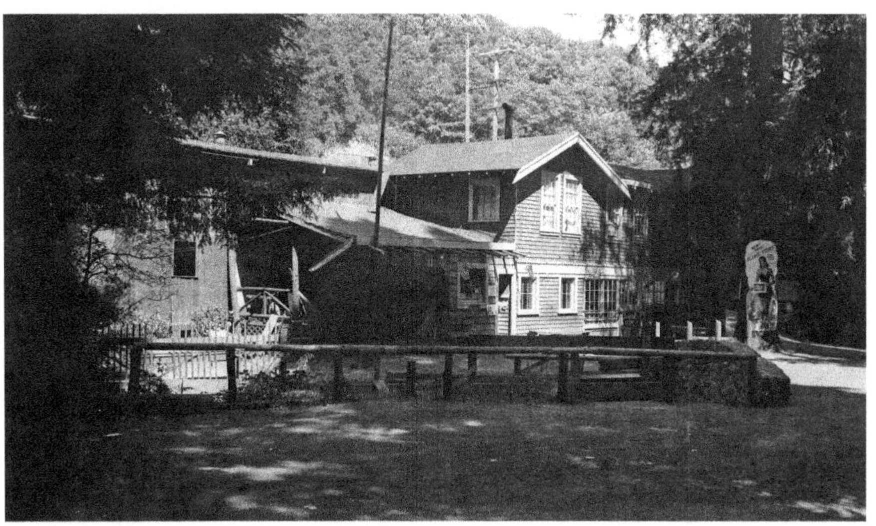

The General Store and dance hall was the community center of Canyon from the 1920s until it burned down in 1969 after a nearby pipeline carrying jet fuel was dynamited. *Courtesy Western Railroad Museum.*

The Canyon Water Brothers Is Formed

On the west side of Canyon, a group of fourteen people organized to buy twenty-seven acres of land. Each contributed several thousand dollars. They called themselves the Water Brothers.[193] The name was taken from the symbolic practice where people formed bonds through sharing water, as described in the popular science fiction book *Stranger in a Strange Land* by Robert Heinlein.

According to Van Der Zee, Canyon Water Brothers became a legal association on September 20, 1967:

> *What you actually got for a $5000 minimum investment in the Water Brothers was an unimproved home site on which you were not legally allowed to build anything, the possibility of renting space in one of five run-down summer houses, and the chance to live in proximity to a bunch of people you hardly knew, in a community in danger of going down entirely. It wasn't really an investment at all, but a declaration of faith.*
>
> [T]*he Water Brothers drew...extraordinary investors: a schoolteacher, a lawyer, a tax investigator for the government, a civil engineer with a*

PhD—some of them complete strangers to Canyon and to the people who lived there. In this manner, Mrs. Holmes' price was met in full and in cash, within the allotted time, and twenty-six acres of Canyon land that East Bay MUD badly wanted remained in the hands of the community.

The Canyon people who had organized the Store Trust had used their own property as a base and had been able to raise the final $4000 with a bank loan. The new group, however, had almost no land or houses and had no prospects of bank assistance. They would have to put up either their own cash, or whatever they could raise without collateral.

[Those] *who started the Canyon Water Brothers offered $60,000 that they did not have. They raised the funds, and Mrs. Holmes, the store owner, agreed to sell to the residents, rather than to the utility.*

Contra Costa Health Department immediately condemned the building. Canyon Construction Company and the Store Trust applied for a building permit for a new store containing a residence, an assembly hall and a new post office building. EBMUD objected because a septic tank would be the only feasible way to handle sanitation.

Over Thanksgiving weekend of 1967, Canyon Construction led residents in rebuilding the General Store, even jacking up the structure and placing a new concrete foundation under it. Women served food to the men. Children worked along with the men in the crew. Complete strangers stopped by to lend a hand. By Monday morning, the building had been largely restored. The county issued a stop-work notice. Negotiations continued. But before the matter could come before a public hearing, disaster struck.

The Jet Fuel Pipeline Explosion

Although few were aware of it, a pipeline carrying aircraft and automobile fuel crossed San Leandro Creek upstream from Canyon. It had been installed in 1965 to pump the volatile fuel from the Shell Oil refinery in Martinez to storage tanks in Oakland and San Jose.

Early in 1969, a labor dispute broke out in the Shell Oil western regional division. Employees went on strike and called for a consumer boycott. The strikers stayed out for ten weeks, returning to work without a contract on March 17, 1969. That night, someone dynamited the pipeline over San Leandro Creek.

At the Shell plant in Martinez, workers had just finished pumping a large volume of gasoline to San Jose. Although the pumps had been shut down, there was still fuel in the pipeline. Suddenly, a pressure gauge plunged, indicating a leak on the line. Earl Davis, a pipeline superintendent, scrambled to locate the rupture. Van Der Zee related the story:

> *Meanwhile, in Canyon, thousands of gallons of Shell gasoline were gushing out of the broken pipe into San Leandro Creek, where the current carried the fuel and its thick fumes downstream, past the newly renovated store and post office and in the direction of the old Canyon School. People in the community, startled by the concussive thump of the dynamite blast, hurried down the hillside to the area around the store…only to find themselves nauseated by the thick and clinging odor of gasoline. The fumes, heavier than air, would not rise, but lingered instead near the windless floor of the canyon, creating a serious fire hazard….Frightened, the residents nearest the store called the Moraga Fire Department and the Contra Costa County Sheriff's office.*
>
> *By the time the firemen and deputies arrived, people were fainting and vomiting from the fumes, and car engines were stalling from lack of oxygen….*
>
> *When* [Earl Davis] *the Shell pipeline man…located the break and saw the size of the rupture and the amount of raw gasoline that was trapped within the narrow creek bed, he hurried up to the store and told the people standing there to run for their lives….He stepped into the phone booth next to the store to call the refinery….*
>
> *Suddenly, in a series of tremendous roars, the whole night was lit with leaping sheets of orange and yellow flame…a wall of flame swept down the creek, consuming everything in its path.*

In the moment of the explosion, Earl Davis was propelled from the melted phone booth, gravely injured. He was transported to St. Francis Hospital burn unit in San Francisco. After making arrangements for his family, the forty-five-year-old Pleasant Hill resident succumbed four days later.[194]

Rick Rivard was one of the seven people injured in the blast when he was a youth living in Canyon. He is now a member of the West Virginia Native American Elders Council and Repatriation Council; he crafts indigenous flutes and plays traditional music. In *Handmade Lives*, he recalled his experience in the Canyon explosion:

> *We'd borrowed Ephraim's van to go to the airport, and it stalled at the post office, because of the fumes, I guess...the gas ignited and everything blew up. I rolled through the van and out the other door and hit the ditch. It looked like everyone was dead—I thought Betty was burned up. The flames were really tall, and I wanted to get out of them. I was already burned pretty bad, feeling really hot. I started up the ridge; I'm sure I was in shock....I kept moving, somehow made it up to Skyline* [Boulevard] *and found a house. They called the police and got me to a hospital. Betty searched for me all night, thought I was dead too, found me wrapped up in bandages.*

The Canyon store, the post office and one home were destroyed. Although the Shell Oil Company temporarily offered a $50,000 reward for information leading to the conviction of the person who dynamited the pipeline, the perpetrator was never identified. Shell Oil soon restored the pipeline to service. Later in 1969, however, the company was forced to shut it down temporarily, after the Wilshire Heights landslide destroyed nineteen homes. The pipeline passed through the geologically unstable region between the Hayward Fault and the hilltop where a Latter-day Saints temple had recently been constructed.

The pipeline was again reactivated. "Back in the 1970s, the city [of Oakland] made $300,000 annually in franchise fees from Shell Oil," noted John Bouey, a retired construction manager for the Canyon Lakes Project in Contra Costa County. "The neighbors were led to believe that it was a water line, but...it carried 1.4 million gallons per day of jet fuel at a pressure of 1440 pounds per square inch."[195]

Bouey recounted the efforts of his neighbors Marge Saunders and Carol Brownell in opposing the reactivation of the pipeline. In 1985, the City of Oakland discontinued the franchise, and Shell Oil decommissioned the pipeline.[196]

Canyon Citizens Prevail Over EBMUD

Beginning in 1970, under new managers, the utility ceased purchasing Canyon real estate and pressuring residents to leave. In the absence of an adversarial public agency, the community flourished.

Not all was idyllic in the 1960s and 1970s, however. Sometimes children were given drugs. There were cases of pedophilia. As adults, some of the children who had been abused later confronted the townspeople to hold them accountable.[197]

Contributors to *Handmade Lives* reported that they kept up their dear friendships with one another and largely repudiated their parents' lifestyle in favor of monogamy and sobriety.

Canyon in the Twenty-First Century

In the twenty-first century, Canyon continues to thrive as a cohesive community. The history of the residents prevailing against EBMUD, the joys of the hippie era and the continuing excellence[198] of the three-room, K-8 Canyon School are a source of civic pride.

While the General Store was never rebuilt to be the same as it once was, the Canyon Post Office serves as the de facto community center and pocket museum.

Canyon Water Brothers lives on as a legal entity, with its sixteen members managing its increasingly valuable real estate holdings.

In 2016, after a lengthy public process, the old Pereira-McCosker land between Canyon and the Huckleberry Botanic Preserve was donated to the regional park district. The park district held a series of public meetings to allow the Canyon residents and other "stakeholders" to express their views on how the new park should be managed. With their history of trouble with other agencies, some Canyon residents voiced wariness at the notion of the park district, with its power of eminent domain, as a new neighbor.

The unique spirit of Canyon lives on.

The McCosker and Pereira Families

Descendants of the pioneer families who settled in the hills above the Moraga Redwoods after the sawmill era remain local residents.

Pioneer Families of Canyon, California

Patrick and Catherine McCosker were Irish immigrants who homesteaded on the north side of Thorn's (Pinehurst) Road, starting around 1864. Canyon at that time was populated by about one hundred residents who remained after the Thorn mill had closed.

Joseph Machado Pereira emigrated from Portugal after making four journeys as a crewmember on a Portuguese ship to the New World. He and his wife, Aemelia (Maria) Silviera, also homesteaded on 160 acres in the Canyon area.

Robert Manes, an Irish immigrant and a friend of the Pereiras, homesteaded 160 acres adjacent to the Pereira homestead. In the 1930s, the Pereira family acquired the Manes acreage, giving them 320 acres. The land was used mainly for raising cattle for market but also for growing fruits and vegetables for family consumption. Some dry land farming of oats and barley for winter feed also took place.

These homesteads were part of Joaquín Moraga's Rancho Laguna de los Palos Colorados. The lands are now part of Huckleberry Botanical Regional Preserve.

Patrick and Catherine McCosker's oldest child, John, and Joseph and Maria Pereira's oldest child, Mary, married and moved onto 40 acres that John had homesteaded in 1887 adjacent to the Pereiras' 320 acres.

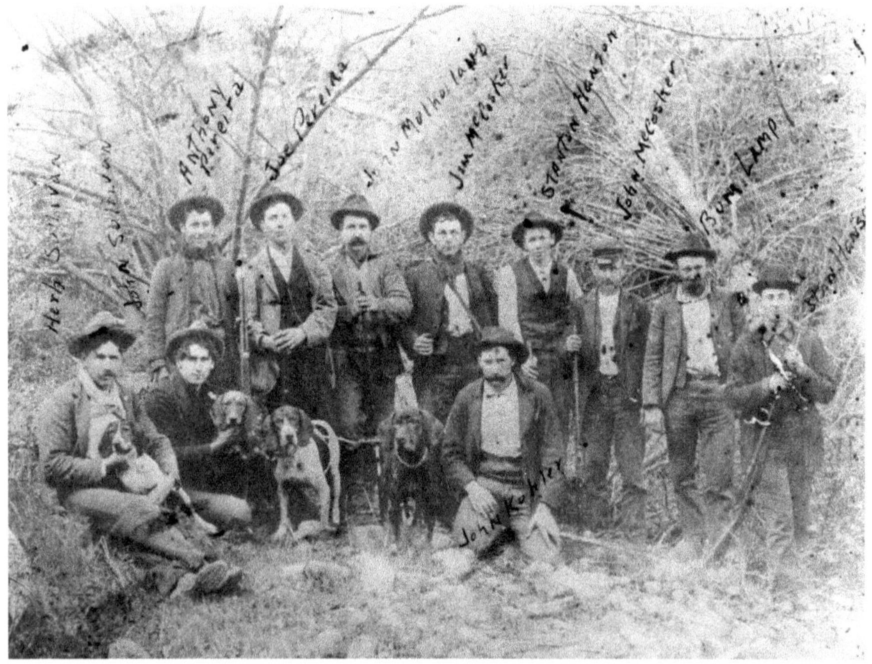

Founders of the town of Canyon include members of the Pereira and McCosker extended families. *Courtesy Moraga Historical Society.*

All of John and Mary's children remained in the East Bay. "My father, Alfred J. McCosker, was the first graduate of Canyon School," recalled Dwayne McCosker.

> *He was the only pupil in the class of 1922. He and his siblings rode the Sacramento Northern Railway into Oakland to attend Oakland Tech High School. Dad and his siblings would board at Wilcox Station, which was at what is now John McCosker Ranch Road, travel past Eastport Station, through the tunnel into Oakland.*
>
> *During the 1960s, the State of California planned to construct the Shepherd Canyon Freeway, Highway 77. It would have come from the Warren Freeway (Highway 13) in Montclair through Shepherd Canyon, the Pereira Ranch, the Domingo Ranch, along Adobe Ridge on the Silva Ranch, through Moraga and eventually, I believe, was to connect to what is now the 680/24 interchange in the vicinity of Pleasant Hill Road. Even though it was preliminarily designed and staked, it was never built.*

Uncle Joe Pereira's Stories

Dwayne McCosker heard many local stories from his Uncle Joe:

> *I enjoyed going to Uncle Joe and Aunt Cora's. As Dad's uncle and aunt, they were my great-uncle and -aunt. Joe always had good stories to tell, even if they were told time and time again.... One time I mentioned the famous bandit Joaquin Murietta to Uncle Joe. "Oh, he wasn't such a bad guy," said Uncle Joe, who went on to tell the following story.*
>
> *Once in the early days, Joe's father, the pioneer Joseph M. Pereira, was robbed of $100. Uncle Joe remarked that $100 "was a lot of money in those days." The bandit told Grandfather Joe that he was Joaquin Murietta. "I need to get out of this country in a hurry," said the bandit. "You will be paid back." Of course, Grandfather Joe did not believe this. One day, years later, Grandfather Joe and his family heard a commotion at their ranch house. The dogs were barking, but no one could see any intruder or anything going on. When they looked on the front porch, they found a sack containing $300. Grandfather Joe could only interpret these strange doings as the bandit having returned to repay his debt.*

Sharing the Ranch Land in Perpetuity

"After Alfred [A.J.] died in 1987 at the age of 78, and Aileen died in 1992 at 82, the family decided to sell the ranch land, with the exception of the 40 acre Home Ranch which John McCosker homesteaded, beginning in 1887," Dwayne McCosker explained. "The ranch was sold to Montenara/Indian Valley Land Company. They have used it as mitigation property which has been gifted to public entities, in order for Montenara to obtain permission to develop their Wilder subdivision. Most of the original Pereira property is now part of Sibley Botanical Reserve. The remainder of the McCosker Land and Cattle Company Canyon Ranch is now owned by East Bay Municipal Utility District as watershed."

The Sacramento Northern Railway

The Sacramento Northern Railway (SNR) was an interurban electric commuter trolley that, remarkably, carried passengers 183 miles from Oakland to Chico in the years between 1918 and 1941. In an era when most rural roads were unpaved and could not handle many automobiles, the train provided a vital connection for residents of the hidden canyons of the East Bay redwood region.

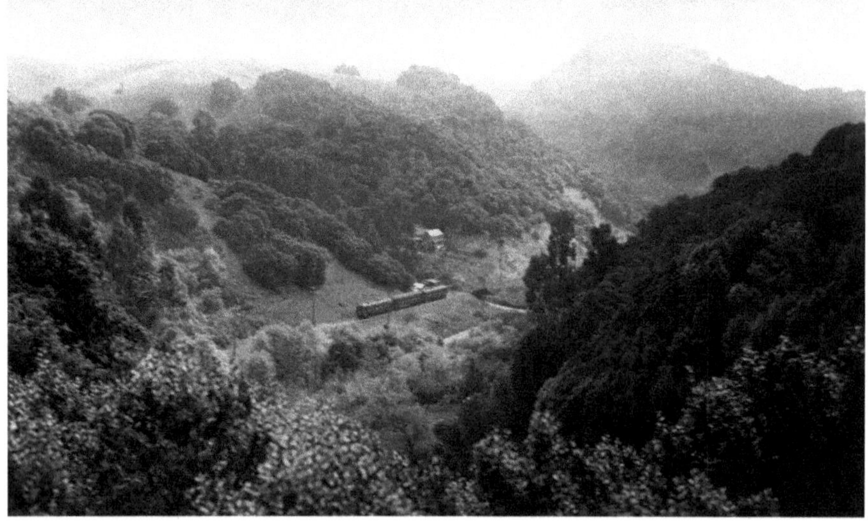

The Sacramento Northern Railway wends its way through Canyon on its way toward Chico. *Courtesy Western Railroad Museum.*

Few aspects of Oakland history have so captivated a small but enthusiastic coterie of train buffs—notably the late authors Vernon Sappers and Harre Demoro. Part of the allure stems from personal recollections, still told by a few people, of their experiences riding or watching this train. The east shore of Lake Temescal and the Montclair commercial district are built around the grades and massive concrete buttresses along which the trains once passed.

"We would pick up the Sacramento Northern across from the Valle Vista staging area and ride it out to the Little Pond Restaurant in Moraga. They had champagne breakfasts," Lloyd Graham Donaldson recalled. "There was an old ranch behind Valle Vista, at the back side of the dam. We'd ride the train to Moraga from there, have lunch and ride back in."[199]

The East Bay Hills Project website, inaugurated in 2011, showcases a gallery of Sacramento Northern photos.[200]

Routes

The Sacramento Northern was incorporated in 1918 when two separate railroads merged. On the southern end, the Oakland, Antioch and Eastern Railway, starting in 1913, ran between Sacramento and the Oakland pier terminal (the "Mole") operated by the Key System of East Bay streetcars.

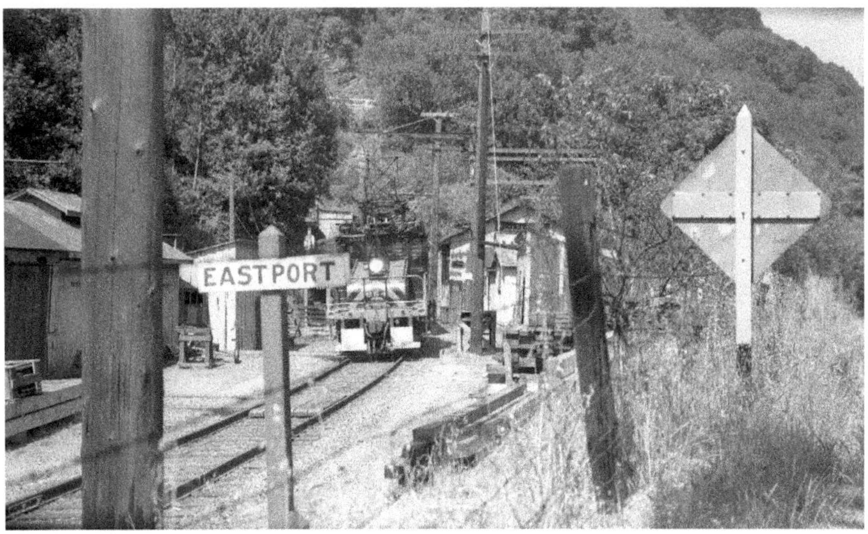

Northbound trains emerged from the Redwood Peak tunnel at the whistle stop called Eastport. *Courtesy Western Railroad Museum.*

Beyond Sacramento, the Northern Electric Railway, first constructed in 1904, followed the eponymous river to arrive at Chico.

At Fortieth and Shafter Streets in Oakland, the SNR had a small "Oakland Yard" that marked the point where westbound trains transitioned to the Key System tracks and propulsion system. From this station, eastbound trains climbed the relatively steep grade[201] past Lake Temescal, through the Montclair district and up Shepherd Canyon. The passage through the Oakland hills to Contra Costa County was through what the railroad called the Redwood Peak Tunnel.

Trains emerged from the tunnel at Eastport Station. Along a more gradual slope along the south side of an upper tributary of San Leandro Creek, there were additional stops called Wilcox[202] and Canyon.

At "the crossroads," where today's Pinehurst Boulevard makes a "T" with Canyon Road into Moraga, there were stops at Pinehurst and Valle Vista. The railway continued on to Walnut Creek, Concord, Pittsburg and Sacramento.

At Pittsburg, entire passenger trains rolled onto a ferryboat called the *Ramon*, which carried them across Suisun Bay to a landing on Chipps Island.

History of Operation

During its heyday, the Sacramento Northern adhered to first-class passenger train standards, with parlor/observation and dining cars. Named cars included the "Bidwell," "Sacramento," "Moraga" and "Alabama." The latter had been custom-built in 1905 for railroad tycoon Henry E. Huntington. The "Alabama" was purchased in 1921. Its passengers enjoyed fine dining therein until 1931, when the "Alabama" was destroyed after its electric coffeemaker caught fire.[203]

Trains with names included the Comet, Meteor, Sacramento Valley Limited and Steamer Special.[204] In 1939, SN operated three weekday trains from Sacramento to San Francisco. Traveling at speeds up to sixty miles per hour on the rural stretches, the fastest train trip took five hours and forty-three minutes from Oakland to Chico.

The Sunday Picnic Specials and the Canyon Parks, 1913–41

In order to encourage ridership—and, not coincidentally, bring prospective buyers to see Canyon real estate—the railroad owners constructed and maintained three amusement parks. Between April and October, the railway company chartered picnic trains that went to the three parks in the fragrant redwood groves: Canyon, Madrone and Pinehurst Park.

Janet McEwan, in her unpublished thesis on the history of Canyon, discussed the role of the picnic trains:

> *There followed the railroad's biggest years, when for only a dollar city people could get away to the country for a picnic or a day of hiking. Canyon, as a scenic spot, also enjoyed the height of its popularity.*
>
> *The "Pinehurst Special" would be advertised and entire trains would be chartered by groups such as the Native Sons of the Golden West or the Knights of Columbus, A round trip ticket plus park admission cost one dollar in 1925 for adults, 25 cents for children. Park admission cost fifty cents for those who hiked in.*

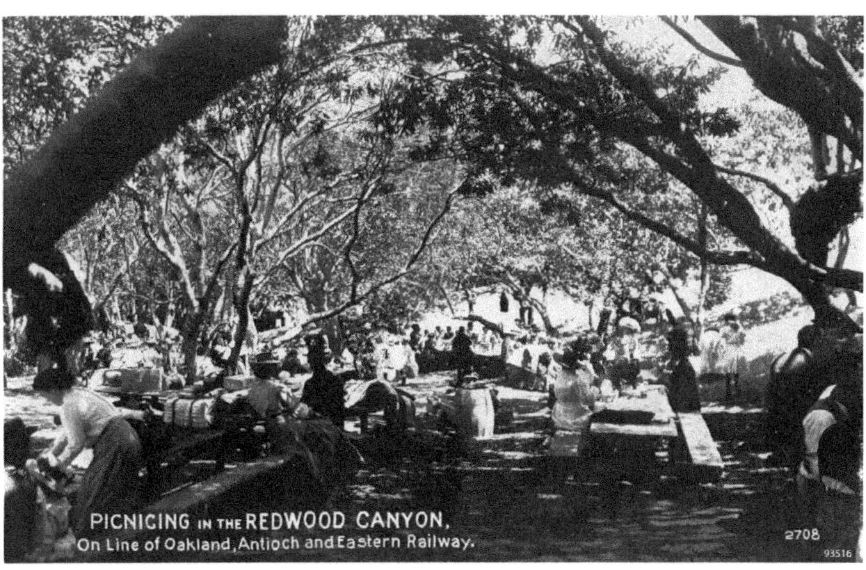

Railroad managers advertised three amusement parks at Sacramento Northern stops. Pinehurst had a large picnic area, bandstand and dance floor. *Courtesy Western Railroad Museum.*

Trains consisting of from four to ten open picnic cars left the 40th and Shafter station to arrive at Canyon station in 20 minutes, and at Pinehurst station five minutes later. Often during the peak season, extra cars had to be rented from the Key System and adapted for proper coupling with the Oakland-Antioch picnic cars.[205]

THE AMUSEMENT PARKS OF CANYON

Across the road from the Redwood Inn, which was located near the original Canyon general store and post office, was Canyon Park. It featured a merry-go-round. Albert McCosker was at one point its "conductor."[206]

Pinehurst and Madrone Parks were located at the "crossroads," where Pinehurst Road meets Canyon Road into Moraga. The exact locations of these parks are not known, but both featured picnic tables. Pinehurst Park also had a dance platform, a bandstand and a baseball diamond.[207]

It is not clear whether the railroad owned the land on which the parks were located. The railroad probably leased the land originally, but it maintained the parks and paid the electric bill for the merry-go-round. The sale of Pinehurst Park to the railroad in 1920 is documented.[208] Pinehurst and Madrone Parks would appear to have been built on land originally owned by Duncan McNee, and later by Harry A. Mitchell, that was then sold to the East Bay Water Company.[209]

THE ROLE OF THE SACRAMENTO NORTHERN RAILWAY IN REAL ESTATE DEVELOPMENT

Frank C. Havens corresponded with other businessmen, starting in 1902, regarding how building a railroad to link Alameda and Contra Costa Counties would increase his property values east of the hills. Shortly thereafter, he acquired the rights to an electric railway for all of Contra Costa County.[210]

Once the train was providing service, the real estate boom commenced. Picnic train passengers were met at the station by real estate agents. According to McEwan:

These picnic excursions provided valuable exposure for the subdivisions of such real estate men as Ellis Wood and his agents. Wood's real estate

> *office was located...near the Canyon station....Other* [realtors] *sold land through advertisements in the* Oakland Tribune, *with the average lot priced at about $1,500. This kind of advertising was only profitable after the inception of the successful picnic excursions. Before the railroad made Canyon accessible to most city dwellers, there was little market for summer cottages.*

The Railroad Subdivisions

On the west side of the Redwood Peak Tunnel, Havens and the Realty Syndicate had set to work subdividing land and trying to sell lots in Shepherd Canyon in the early years of the twentieth century. With the construction of the Redwood Peak tunnel for the Oakland, Antioch and Eastern Railway,, Havens constructed a station building there to promote his subdivision.

The station stop was called Havens. Its passenger shelter, located near today's Paso Robles Drive and Shepherd Canyon Road (then called Park Boulevard), was of a mission style unique to the Sacramento Northern Railway but similar to other shelters used for the Key Route streetcars.

Havens Station, however, served few passengers, and even fewer lots were sold. This may have been due to the remote location, five miles uphill from the Oakland Yard at Fortieth and Shafter. Also, the flat land in the canyon was all very close to the railroad right-of-way. Only a few houses were built on the steep hillside.[211]

The Realty Syndicate went bankrupt in 1915. Meanwhile, on the other side of the Redwood Peak railroad tunnel, land sales were much more successful. Uphill from the town of Canyon, formerly the haunts of bullwhackers and saloon patrons, the largest and most successful of the "railroad subdivisions" was Moraga Redwood Heights. The land was purchased by speculator George W. Austen in 1911 and sold in 1914 to Ellis and Minnie Wood.[212]

Moraga Redwood Heights lots initially, from 1912 to 1915, were sold to prosperous East Bay or San Francisco doctors, lawyers and corporate presidents for weekend or summer homes. After 1914, more land was subdivided, and lots were sold to teachers, mechanics, small business owners and skilled workers such as stenographers. The initial subdivision called for a separation between residential and commercial districts. Lots came with deed restrictions prohibiting the sale of liquor and the operation of public stables or garages. Deed restrictions also prohibited resale to any non-white persons.

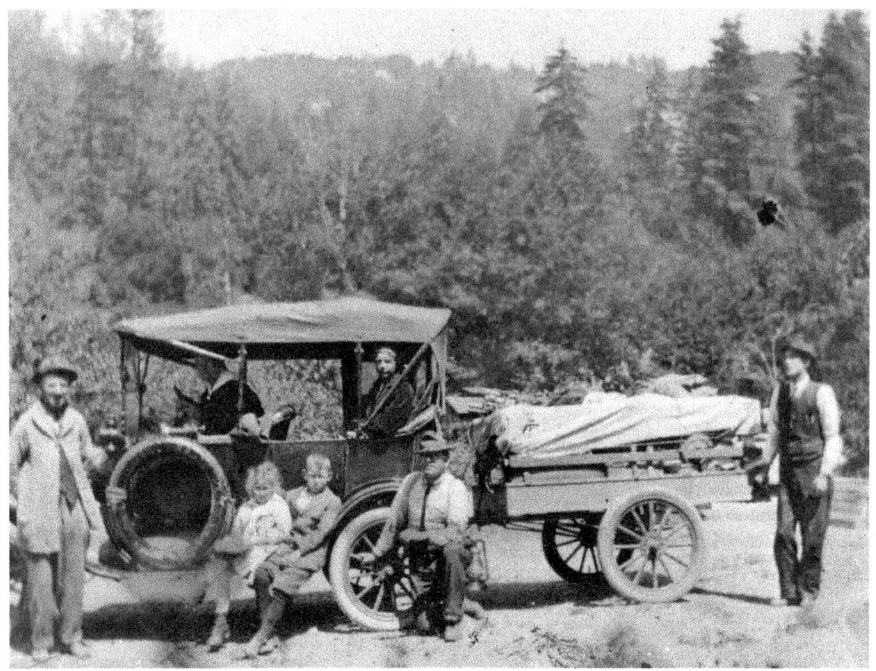

The Moraga Redwoods, with hilltop vistas, were attractive to families wanting to camp in the wilderness above the town of Canyon. *Courtesy Moraga Historical Society.*

The success of Moraga Redwood Heights encouraged other developers. Farther down the canyon toward Pinehurst, Wilcox Uplands and Alder Dell were constructed. These were located farther from the scenic redwoods surrounding canyon and less successful.

Alder Dell was subdivided by George Lydiksen, a building contractor. Despite this background, Lydiksen surveyed the lots incorrectly. Overlapping lot lines proved to be a source of confusion and some litigation for the next fifty years and beyond.

Land sales dropped sharply during the wartime economy of 1917–18. Thereafter, another building contractor, Woodward J. Martinez, who also owned an electrical supply business in Berkeley, filed a subdivision map for Canyon City in 1919 in partnership with Ellis Wood.

Canyon City, located upslope from the Redwood Inn and Canyon commercial district, was more successful. Most of the lots had been sold by 1929.

Also in the 1920s, a subdivision called Moraga Redwood Heights Extension was opened, adjacent to its upscale original namesake. It was developed by the Wickham Havens Villa Site and Development Company, on land that had been purchased by Havens's Mahogany Eucalyptus and Land Company. Sales peaked in 1924 and had ceased by 1930.

As McEwan noted, "The twenties were a period of steady growth and solidification of a middle-class eithic...[by then] Canyon had begun to identify itself as a community. The institutions by which such an identify is molded were created primarily in that decade."[213]

"These developments consisted mostly of summer cottages, built individually to fit their surroundings—small, comfortable wooden houses with large porches to take advantage of views, trees, and sunlight, according to Van Der Zee. "The steepness of the hillside made the cutting and filling necessary for elaborate construction prohibitively expensive, and the scarcity of water placed and absolute limit on population density. By the mid-1920s, after 110 homes had been built, development was halted."[214]

Passengers versus Freight

With the construction of the Bay Bridge in 1939, passenger service was extended into San Francisco. But carrying people proved to be a money-losing proposition for the owners, the Western Pacific Railroad Company. Ridership had decreased during the Great Depression and with the increasing popularity of automobiles. Passenger service was discontinued altogether in 1941.

Short-haul freight service, however, was economically viable. During and after the World War II years, Sacramento Northern trains were used to carry war materiel. Dwayne McCosker recalled his fascination with the passing SNR trains when he was a student at Montclair Elementary School around 1949: "My teacher, Miss Head, would continue to speak, even while the train was passing by the classroom window carrying tanks, howitzers, jeeps and trucks. Of course, all the boys and many of the girls wanted to look at the train. Miss Head would close the venetian blinds and just keep on teaching."[215]

The Demise of the Sacramento Northern Railway

Eventually, economics caught up with the Western Pacific Railroad, the SNR parent company. Ridership had declined during the Depression years.

The Caldecott Tunnel was completed in 1937, making it easy for drivers to get through the East Bay hills. Soon the automobile surpassed rail as the more popular mode of transportation. Passenger service west from Sacramento ended in 1940. All passenger service ended in 1941. But freight service continued until 1954.

As Oakland, Los Angeles, and other cities redesigned themselves around the automobile in the late 1950s and early 1960s, light rail track was removed everywhere. Thus it was with the Sacramento Northern Track, even inside the now-unused Redwood Peak tunnel. "My brother and I drove through the tunnel a few times after the tracks were removed, even though we weren't supposed to," recalled Dwayne McCosker. Adventurous children enjoyed walking through, as well. Eventually, the authorities sealed both ends of the tunnel with concrete plugs to prevent accidents.

John Vallerga, a UC-Berkeley scientist, recalled an adventure he and his friends had in the sealed tunnel during the late 1960s:

> *Word got out in Montera Junior High that some kids were trying to dig into the sealed-off tunnel in Shepherd Canyon using picks. A crowd of boys had gathered there when they opened the hole in the concrete. Once it got big enough for a 14-year-old boy to squeeze in, they shoved a ladder through the hole to drop down 10 feet to the bottom of the tunnel.*
>
> *Many of us scrambled down the ladder, and it was pitch black except for the light coming through the hole. Behind us was the concrete wall. In front of us was the tunnel. We decided to walk towards Moraga.*
>
> *Some kid lit an emergency car flare, which was too bright and very smoky. We walked into the tunnel about 100 yards and came upon what looked like a cave-in blocked by rubble. So we turned around and got out, mostly because of the smoke.*
>
> *That was that, I never went back, and I heard that Public Works sealed the hole pretty quickly, this time with rebar.*

The Towns EBMUD Destroyed

Starting in the 1930s, managers at the East Bay Muncipal Utility District worried about the threat of pollution from sewage polluting the drinking water in San Leandro Reservoir.[216] With no municipal sewer systems available in rural Canyon and the surrounding towns, every house was on a septic system.

The water agency developed its sanitation plan: whenever a rural house was listed for sale, the company would purchase the house at market value and demolish it. In some cases, this philosophy extended to buying and demolishing entire towns.

Pinehurst

The site of the small town of Pinehurst is located where the two arms of Pinehurst Road meet Canyon Road, connecting the forest to the town of Moraga. Here, where redwoods tower above the cool and shady creek,[217] was once the site of a Saclan village.[218] Pinehurst was also the site of the first sawmill in the Moraga Redwoods, built by Taylor and Owen in 1849. A small town may have remained at the site before it became a stop on the Sacramento Northern Railway. Local lore mentions a tavern and maybe a bordello.

Valle Vista

Only a mile beyond Pinehurst heading toward Moraga on Canyon Road, the site of Valle Vista[219] is easy to locate. Where the town once stood is a vast paved parking lot, which EBMUD has named the Valle Vista staging area. It serves as the trailhead to Upper San Leandro Reservoir.

A bequest of family photos from one-time Valle Vista resident Wilbur Hespe to the Moraga Historical Society allows us a glimpse of how the

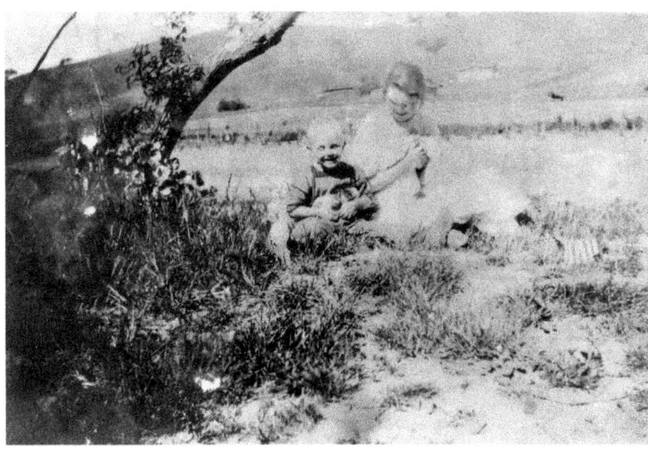

Young Wilbur Hespe and his mother are shown in the vanished town of Valle Vista, circa 1915. *Courtesy Moraga Historical Society.*

town once appeared. Residents had no utilities. They grew vegetables in household gardens. Most residents worked in a nearby rock quarry. On days off, they could board the Sacramento Northern and visit Oakland to the west or Moraga, Walnut Creek and points east. Valle Vista children attended the Canyon School upstream, which had been founded in about 1918.

The town was demolished by EBMUD in the mid-1960s.

Redwood

The town of Redwood was located on San Leandro Creek prior to 1900, when the area was flooded when Lake Chabot filled up behind its dam, creating the Upper and Lower San Leandro Reservoirs upstream.

Local lore[220] recalls that the Redwood School that operated near the entrance to Redwood Regional Park between about 1900 and 1964 was originally in the town of Redwood, deep in the valley of San Leandro Creek. Some of the Portuguese immigrant children would make the long walk to school and back.

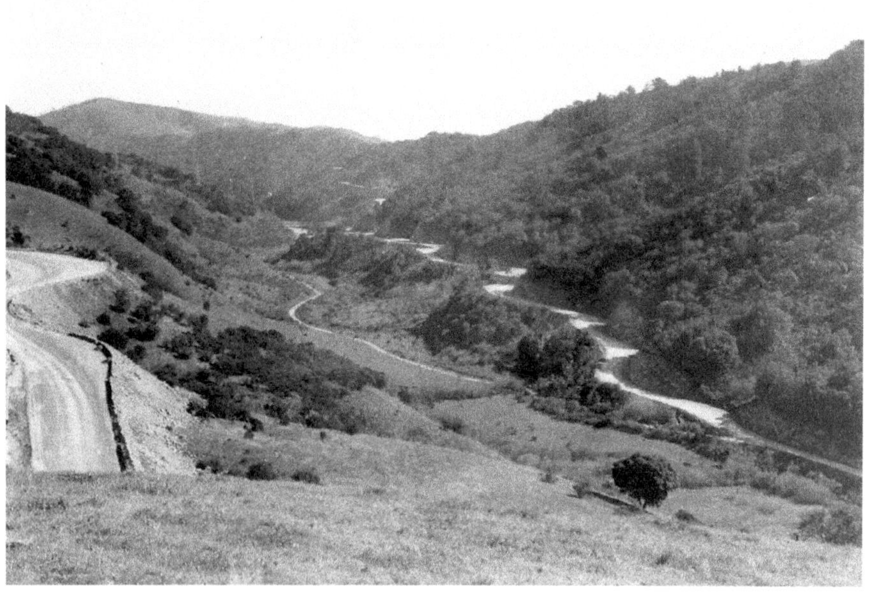

This photo, looking toward Castro Valley after the WPA finished paving Redwood Road, shows where the diverging unpaved road leads to the vanished town of Redwood. *Courtesy East Bay Regional Park District archives.*

During the New Deal years, the Works Progress Administration (WPA) and its state counterpart, the State Emergency Relief Administration (SERA), paved Redwood Road between the regional park main gate and Castro Valley. A 1949 photograph shows an older road diverging from the newly paved road, as viewed looking southwest. It is believed that the older road led to the town of Redwood.

Grass Valley (now Anthony Chabot Regional Park)

Homesteads in Grass Valley

Grass Valley, now part of Anthony Chabot Regional Park, was never known to have been the site for a town. The area was a traditional homeland of the Jalquin-Yrgin people. However, according to a 1982 cultural resource report compiled by Peter Banks for the EBRPD, it is suspected that there were no permanent Native settlements there due to a lack of reliable year-round water in its creek.[221]

The 1982 cultural resources study states that there were five structures in Grass Valley from 1880 to 1934. Two were in the Bort Meadow/Big Trees area. Banks quoted Lloyd Graham Donaldson as saying that these were the house and barn of the Grass Valley Ranch.

Another homesite was at the Stone Bridge, south of Bort Meadow and just north of the Jackson Grade leading to the Woolridge staging area. A structure at this location was shown on the 1947 USGS topographical map. It was undoubtedly demolished by the park district, which sometimes unobtrusively burned down unwanted buildings then, at time when the area was far less developed.

Interestingly, the Banks archaeological report did not discuss the most prominent homestead site, where the Maciel family (later changed to Marciel) lived, where the gun range would later be developed.

Banks suspected that there may have been an additional homestead site at the intersection of the Brandon and Horseshoe Trails, at the closest approach to the Equestrian Center. A row of locust trees was evidently planted many decades ago. From horseback, a visitor can pick and enjoy the fruits that still grow there. Doris Marciel has no information from her family about these trees.[222]

Also mentioned in the Banks report is the John Cahill homestead. This was shown on the 1899 USGS map. He apparently sold the property to the Contra Costa Water Company, and the buildings were demolished before Lake Chabot inundated the area.

Photographs from the 1950s, in the EBRPD archives, show a storefront with a sign that reads, "Grass Valley Post Office." This was a town mockup used as a playhouse for children; the Grass Valley Ranch area never had its own post office.

Cattle Ranching in Grass Valley

Grass Valley Ranch was 525 acres in area in 1915 when it was sold to the Contra Costa Water Company. According to the Banks study, aerial photographs taken in 1935 showed no buildings at the homestead site near the Stone Bridge. A cattle-loading chute and the ruins of a corral can be found on the south side of the gate that separates the Grass Valley Trail from the Bort Meadow group camping area.

The Grass Valley Trail itself was historically used to drive cattle between Oakland and Moraga/Orinda.

Trees in Grass Valley

While Grass Valley was never thickly forested with old-growth redwoods, only a few miles of rolling hills separate the area from the Middle Redwoods to the north.

According to the Banks report, isolated redwoods were once found in Grass Valley as far as the Redwood Golf Course area, at the base of the hills to the south of the park, at the outskirts of Castro Valley. All the trees were logged; second-growth trees are present.

Starting around 1910, the East Bay Water Company and Contra Costa Water Company planted thousands of eucalyptus trees. In the north end of the valley, a eucalyptus grove thrived in area now called Bort Meadow. It was originally named the Big Trees area, after the eucalyptus.

By the 1930s, the duties of the CCC crews at Camp Grass Valley included cutting the dense eucalyptus overgrowth on the shores of Lake Chabot. Management of the fire-prone and prolific eucalyptus has been a challenge for park managers in Anthony Chabot Park into the twenty-first century.

The Marciel Farm

Manuel and Isabel Maciel,[223] pioneers of today's Anthony Chabot Regional Park, came to California in about 1870 from the Azores Islands of Portugal. They farmed acreage in the hills above Castro Valley, extending from Anthony Chabot Regional Park to the vicinity of the Redwood Golf Course. The family leased the property from Anthony Chabot's East Bay Water Company.

Today's Marciel Gate on Redwood Road near the regional park campground is named in honor of this family. Their homestead was located next to the site of the marksmanship range.[224] "The ranch house stood to the left[225] of where targets were at the base of a cliff," explained San Lorenzo historian Doris Marciel. "My cousin Darrell Lopes examined the area with a metal detector and found many artifacts—old horse shoes, tools and other metal objects....They also farmed the golf course area—the land there was perfect for growing potatoes."

The Maciel children attended the one-room school in Redwood Canyon. This required them to walk several miles down the hill and up again after school. Many of their classmates came from Portuguese immigrant families; the children spoke English in school and were bilingual.

Some of the boys rode in the first local rodeo, held at the Burbank School[226] in Hayward. Rancher Harry Rowell was a key figure in establishing this essential western sport in the East Bay. His Castro Valley spread is now a rodeo park operated by the Hayward Area Recreation District.

Doris Marciel shared some of her family history:

> *My grandfather Manuel was born in 1872 and died in July 1920 at the age of forty-eight.*
>
> *One day he was sitting under a tree, eating his lunch, when two young men were in the vicinity, hunting. They saw a movement under the tree, thought it was an animal and shot him. There were no neighbors nearby. The hunters went to the closest neighbor, Mr. Phillips, who took Manuel to Fairmount Hospital. But by then he had died. He is buried at Mount St. Joseph cemetery.*
>
> *My grandmother Isabel was left with twelve children; they'd had fifteen, but only twelve lived to adulthood, eight boys and four girls. The eldest was twenty-two. The youngest was ten months old when her father died.*
>
> *The older children ran the ranch for three years after Manuel died. They left the ranch in 1923. The empty house was burned down later.*

Legends from Castro Valley recall the era when the old-growth redwoods were logged during the 1850s. At that time, the United States army shipped redwood trees to San Francisco. The lumber was carried on horse-drawn wagons from the sawmills, down Redwood Road to Castro Valley. The wagons went down Main Street in San Lorenzo, today's Lewelling Boulevard, to Roberts Landing on the bay. From there, the lumber was shipped to San Francisco. Farm products were shipped as well.

Grass Valley as Parkland

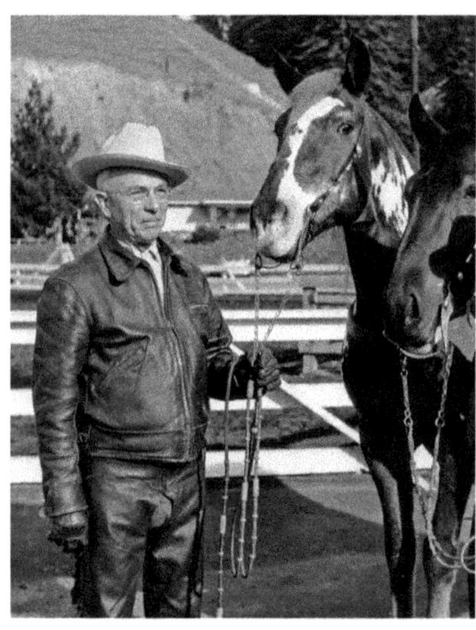

Veteran horseman Claud Brandon rode his pinto gelding, Calico Pete, almost every day in Anthony Chabot Regional Park. The Brandon Trail is named in his honor. *Courtesy East Bay Regional Park District archives.*

The park district purchased Grass Valley from EBMUD in 1952. It was soon opened as parkland. Celebrated horseman Claud Brandon had by then racked up thousands of trail miles with his pinto gelding, Calico Pete. Brandon retired in 1952 and rode from Skyline Ranch to the trails of Grass Valley almost every day until his death in 1967. The Brandon Trail is named in his honor.

The park itself was renamed in 1965 for Anthony Chabot, the "Water King." The Big Trees meadow was renamed for Joseph P. Bort after his death in 1995. Bort was known as a regional transit advocate. He had served as an Alameda County supervisor for the district that included Castro Valley, San Leandro and part of Oakland.

Chapter 7
VISIONARIES

They were planning to subdivide the hills, and I just couldn't let them do it. I started a campaign saying it would be a terrible thing for the Eastbay if they cut up the hills and built houses there. I must have written millions of words about it.
—Harold French

Joaquin and Juanita Miller

Joaquin Miller, the "Poet of the Sierras," was the first visionary to build his castle in the East Bay hills. His bohemian enclave attracted celebrities from the international arts scene. Following his death in 1913, his daughter, Juanita Miller, remained in the aerie they called "the Hights." Enjoying the public acclaim that had been earned by her literary father, Juanita devoted her life to burnishing his legacy.

Tales of Miller's adventures have been carefully sifted by his biographers, as he was well known for his fluid regard for truth. Born in 1837 as Cincinnatus Heine Miller on a farm outside Liberty, Ohio, the poet at age twelve traveled to Oregon on a wagon train with his family. He first fought Native Americans but later went to live with them.

Between 1857 and 1870, he traveled back and forth between Oregon and California. He was introduced to Bret Harte and the attractive librarian Ina Coolbrith in San Francisco in 1870. She persuaded the poet that if he took the name of the bandit Joaquin Murrieta, he would sell more poems.

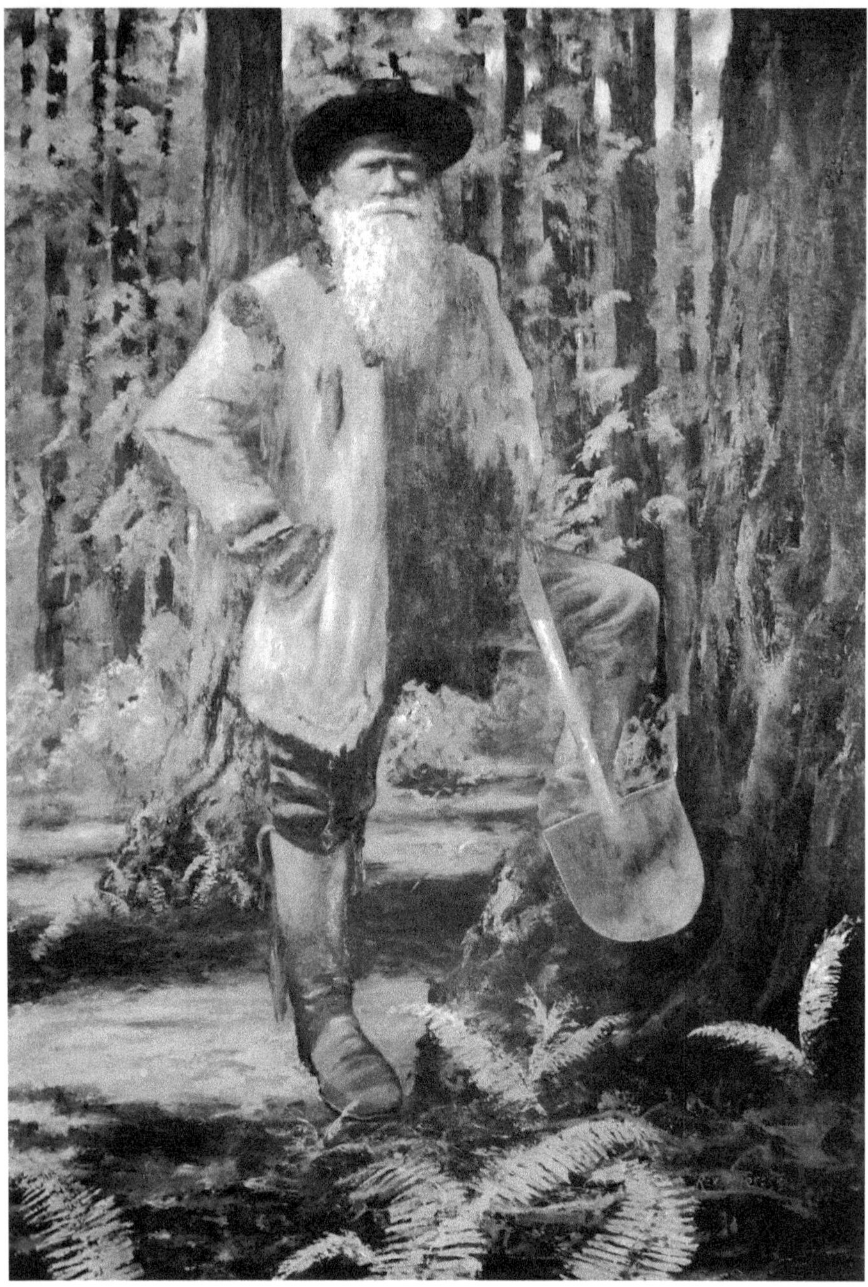

Joaquin Miller, the "Poet of the Sierras," is depicted with a shovel, for his passionate pursuit of tree planting, in this painting by naturalist Rex Burress. *Courtesy Rex Burress.*

Miller soon set sail for London, where he billed himself the "Byron of the Rockies." Sporting a flaming-red shirt, pantaloons, riding chaps, jingling spurs and a *sombrero*, Miller worked the drawing rooms of the aristocracy. He socialized with poet Robert Browning and painter Dante Rossetti.

Having impressed the elites of Europe, Miller returned to America in 1871. In New York, he met Oscar Wilde and actress Lilly Langtry. In the midst of relationships with numerous women, in 1879 he married Abigail Leland, a hotel heiress. Their daughter, Juanita, was born in 1880.[227]

Frustrated by New York City, Miller proposed to Abbie that they go west. "The fastidious Mrs. Miller, accustomed to living in the most luxurious of the Leland hotels, was horrified....She made it plain to her husband that she would not permit Juanita to grow up in a dirty lean-to, and she would not consider living like an aborigine herself."[228] The poet returned alone to San Francisco in 1886.

In his hikes in the East Bay hills, he looked for the site where his hero, John C. Frémont, had gazed at the confluence of San Francisco Bay with the Pacific Ocean and thereby coined the name "Golden Gate." Confident that he had found the site, Joaquin selected it for his homestead.

Dissatisfied with the native grasslands of the Oakland hills, Miller began planting trees. He approached tycoon Adolph Sutro to bankroll this effort. Soon the Hights had become a retreat for literary and artistic luminaries. "He frequently gave 'bandit dinners' on the Hights. He would build a fire outdoors under a big iron kettle filled with venison and onions and a variety of vegetables. During the hours that passed while the stew simmered, Joaquin would talk and recite his poetry and pass around whiskey in tin cups. Later the meal would be served in the same tin cups."[229]

Guests included Jack London, F.M. "Borax" Smith, Frank Norris, Ambrose Bierce and Charles Keeler.[230] Yone Noguchi journeyed from Japan to experience this setting; Joaquin treated the young poet like his valet. Nonetheless, Noguchi was followed by artistic friends from Japan, who built rice paper houses at the Hights.

At some point, Abbie Miller was persuaded to bring Juanita to the East Bay hills. Soon the heiress learned that Joaquin expected her to cook for the bandit dinners. She was not pleased. When news arrived that Abbie's mother was ill, she promptly decamped for New York with ten-year-old Juanita. Joaquin was not to see his daughter again until she was nineteen, in 1899.

Still, the idyllic years of childhood spent at the Hights with her father's artistic circle made a profound impression on the young Juanita.

Joaquin's Monuments

Joaquin constructed—or persuaded his friends to construct—a number of stone monuments on the hillside, structures that remain an enduring legacy to puzzle unwary visitors. The first structure, built in 1892, was a pyramid dedicated to Moses. Next, in 1894, was a phallic tower dedicated to Robert Browning. Tablets, located in what is now the backstage parking lot for the Woodminster Theater, were built in 1904 to honor John C. Frémont.

After clearing a pathway to the promontory that overlooks the Fruitvale district, Joaquin in 1906 constructed a funeral pyre for himself. At the time of his demise, city authorities would not permit it to be used in the manner the poet intended. Local lore holds that some of his ashes were scattered in the vicinity.[231]

Summoning Juanita

In his final months of life, Joaquin summoned Juanita from New York. She came, cared for him in his final illness and stayed at the Hights.

In the early 1920s, Juanita took a screen test and attempted a career in Hollywood, to act in early western movies. No known roles resulted. Still, Juanita found many creative pursuits at the Hights. She published booklets about her father, designed a heart-shaped harp, played the guitar and sang.

In 1942, she commissioned an equestrian statue of her father by sculptor Kise Beeck. It stands at the site of his old corral.

Juanita performed for small groups on the stage at the Oakland Main Library, for which her father's friend Ina Coolbrith had been the first librarian. She spoke to classes at nearby Joaquin Miller Elementary School. "When I was a child, we would go to visit Juanita," recalled longtime neighborhood resident Polly Morris. "She wanted the neighborhood children to know who her father was. She had constructed an effigy of him sitting in his bed, writing. It was creepy!"

Joaquin Miller Days at Woodminster Theater

Juanita became even more prominent after the Woodminster Amphitheater was built in the 1930s by the WPA. From the late 1940s through 1968, each year she produced a theatrical extravaganza called *Joaquin Miller Days*. Local business people and their families took an active interest in participating.

In 1968, the city began contracting with Producers Associates, run by Harriet and Jim Schlader, to produce musicals every summer. This program has been a beloved tradition ever since.

Rangers Assist Juanita in Her Declining Years

After 1968, Juanita suffered declining health. Rangers Rich Wirkkala and Louis (Digger) O'Dell helped her with household matters and brought her groceries. She died in 1970.

VANGUARD OF THE PARKLAND ADVOCATES

Following the death of John Muir in 1914, his friends resolved to carry his environmentalist mission forward. They founded the Sierra Club. Its mission combined public advocacy for nature and open space with an ongoing program of hiking trips and social events. Among the Sierra Club founders were people who would play a key role in preserving open space in the East Bay hills.

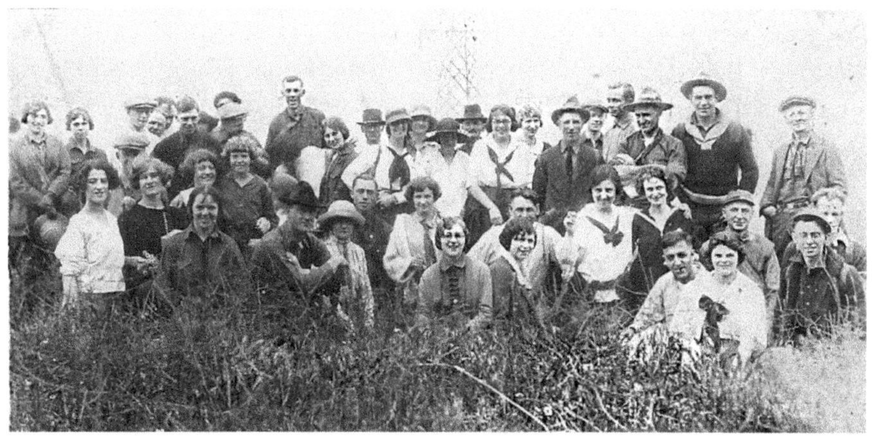

Harold French and members of the Contra Costa Hills Club hiked to Round Top on Sunday, August 2, 1934. French is in the front row wearing a black hat. *Courtesy Contra Costa Hills Club archives.*

Harold French and the Contra Costa Hills Club

Harold French (1879–1962) was a native San Franciscan with a lifelong passion for wilderness. He began his career in journalism as a correspondent from Alaska in 1898, reporting on the Klondike Gold Rush for the *San Francisco Call Bulletin* for two years.

Returning to California, he married Isabel Borthwick in 1901.[232] The couple attended the University of California–Berkeley, where Harold studied mining and Isabel studied English. At the university, they met fellow students and faculty members who shared their great love of mountaineering.

Following graduation, Harold went to work for the United States Mint in San Francisco, commuting on the ferry from their home in Oakland. In addition to this day job, French led hiking parties on weekends and wrote a column called "Hikers" in the *San Francisco Call Bulletin* from 1909 to 1923.

French sought to popularize the wilderness ethic, as expounded by John Muir. Starting in 1912, French publicized the Tamalpais Conservation Club, leading to the preservation of much of the foggy mountain for parkland.

After the death of his mentor, French assumed an active role in the Sierra Club. Later, he turned his attention to his home turf in the East Bay. On February 22, 1920, he organized the Contra Costa Hills Club, with a similar mission to its progenitor, the Sierra Club.

Throughout their lives, Harold and Isabel French led group hikes. He advocated for the creation of Mount Diablo State Park. In addition to his column, he wrote about hiking and open space for the *San Francisco Chronicle*, the *Overland Monthly*, publications of the California Alpine Club and elsewhere. Through his writing, French energized a campaign to preserve Oakland's "Sequoia Park," a large section of which became today's Joaquin Miller Park.

Significantly, French was an early and prominent promoter of the East Bay Regional Park District. Under the rubric of the Contra Costa Hills Club, French sent letters to East Bay Municipal Utility District officials, urging them to relinquish their watershed lands to be used as parks. The campaign proved successful. The French Trail, in Redwood Regional Park, was named in honor of this unsung leader of the early twentieth-century conservation movement.[233]

The Contra Costa Hills Club continues to lead local hikes. The website address is www.contracostahills.org.

The Berkeley Hiking Enthusiasts

By the time John Muir arrived penniless in California in 1868, the College of California had moved from its original site at the Mansion House hotel in downtown Oakland to its permanent location on Strawberry Creek in Berkeley. Soon a group of faculty members and students at the University of California formed to explore the Sierras. At the end of the spring term in 1870, a group of geology students invited their esteemed professor, Joseph LeConte Sr., on a trip to Yosemite. John Muir went along.[234] According to the official history of the Sierra Club:

> *A group at the University of California, led by J. Henry Senger, was interested in promoting recreation by making the Sierra—and especially the Yosemite region—more accessible and better known. Muir joined these and others in the San Francisco Bay Area who were interested in creating an alpine club. Among the organizers were the artist William Keith, attorney Warren Olney, professors Joseph LeConte Senior, J. Henry Senger, and Cornelius Beach Bradley, and Stanford University President David Starr Jordan. Olney and Senger drew up articles of incorporation.*

Thus the culture of enthusiasm for backpacking and mountaineering began around the university community.

Duncan McDuffie and "Little Joe" LeConte

In the 1920s, a new generation of academics and civic-minded business people took up the campaign to open the East Bay hills for public parkland.

Mechanical engineering professor Joseph Nisbet ("Little Joe") LeConte had inherited his father's enthusiasm for mountain packing adventures and for the Sierra Club. He befriended real estate developer Duncan McDuffie, whose philosophy of designing neighborhood streets on natural contours, with abundant trees and landscaping, shaped the aesthetics of Berkeley.

McDuffie camped with "Little Joe" LeConte for one month each summer over more than thirty years. In the process of blazing the segment of the John Muir Trail from Yosemite to Mount Whitney, the two discussed their shared wilderness ethic and the need for parks.[235]

McDuffie and his Sierra Club friends made several first ascents of Sierra peaks. Along with J.S. Hutchinson and J.N. LeConte, he undertook a three-

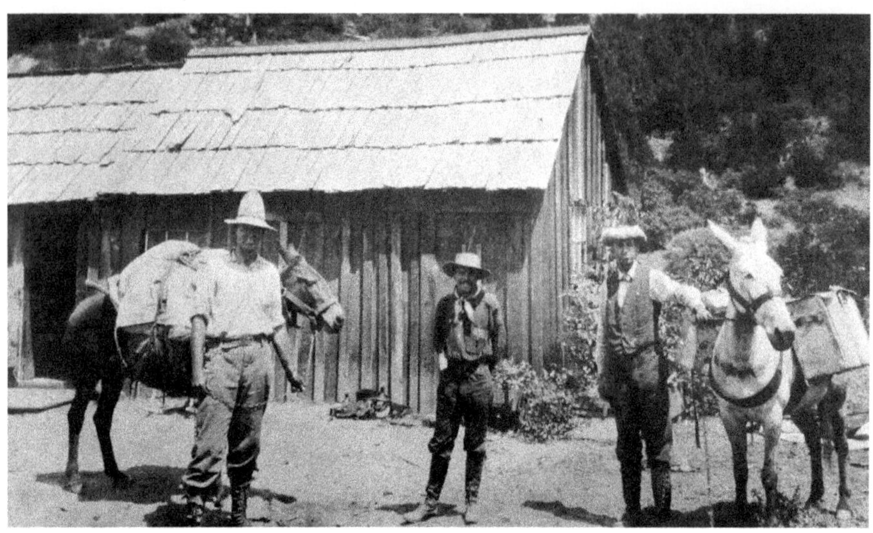

Park advocates who blazed part of the John Muir Trail on a 1908 pack trip to the High Sierra. *Left to right*: real estate developer Duncan McDuffie, UC-Berkeley professor Joseph N. ("Little Joe") LeConte and James Hutchinson. *Courtesy East Bay Regional Park District archives.*

hundred-mile trek over unscouted territory from Tuolumne to Evolution Basin and Muir Pass to the Kings River, culminating in the first ascent of Mount Abbott.

McDuffie went on to chair the Save the Redwoods League. In 1927, he lobbied the legislature to create a California State Parks system. Next, he chaired a statewide committee to advocate for a $6 million bond measure to fund parkland acquisitions. He served as Sierra Club president from 1928 to 1931. He joined the advocates who started the East Bay Regional Park District. He resumed the office of Sierra Club president from 1943 to 1946.

Aurelia Reinhardt

Within a conservationist community that was then predominantly male, the activist president of Mills College played a significant role.

Aurelia Henry Reinhardt (1877–1948) is best known for her role as president of the original Oakland women's college from 1917 to 1943. Reinhardt commissioned architect Julia Morgan to build *Il Campanile.* She supported a vibrant equestrian program on campus, led by Cornelia

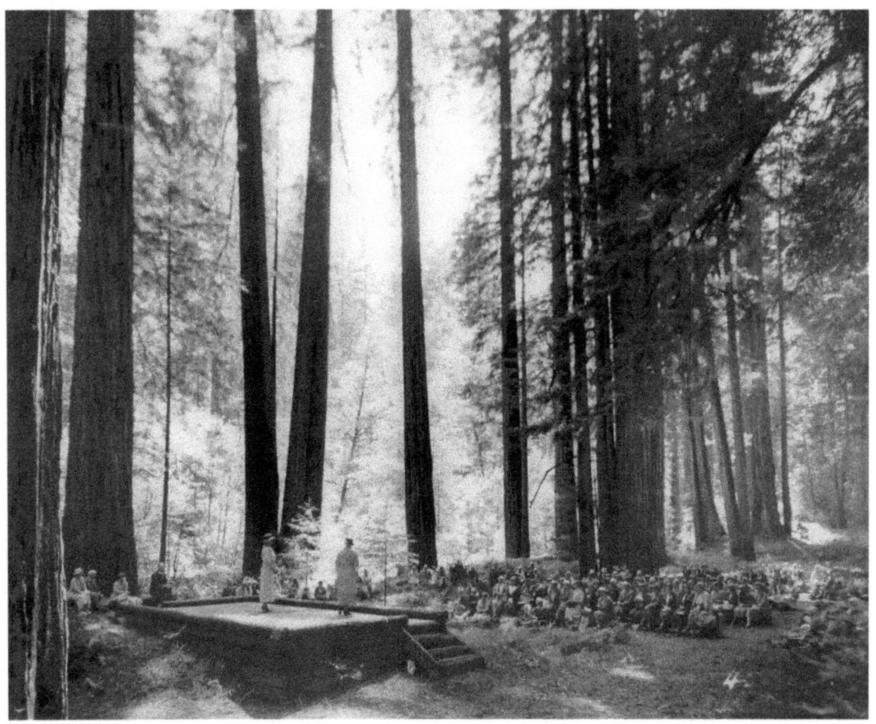

Mills College president Aurelia Reinhardt was an ardent supporter of outdoor recreation and redwood preservation. *Courtesy East Bay Regional Park District archives.*

Van Ness Cress from 1928 through 1955. As college president, Reinhardt required every student at Mills to walk at least one mile each day on the leafy campus.[236] She was also active in the Save the Redwoods League.

Oakland Envisions a Chain of Parks

Early advocates for the East Bay Regional Park District were not the first people to covet the San Antonio Redwoods for parkland.

By 1922, open space enthusiasts within the city of Oakland had envisioned a "chain of parks" extending from Dimond Canyon, through Joaquin Miller's Hights and into the redwood forest. To publicize the issue, members of women's clubs were escorted to Joaquin Miller Park, which then consisted only of the area close to the Hights. Landscape architect Howard Gilkey led the ladies on a hike.[237]

Civic organizations approached the city council and Mayor John L. Davie to request that the city purchase 960 acres of parkland. These groups included the Improvement Clubs of Dimond, Glenview, Brockhurst, Leona Heights and Vernon-Rockridge. The Pioneer Women of Oakland, the Soroptomists and the Fruitvale Women's Club weighed in. The Rotary and Contra Costa Hills Clubs were in favor as well, along with the Melrose and Claremont PTAs. Louis Allen enthused in the *Oakland Tribune*:

> *Instead of the journey of hundreds of miles to the Sierra groves, Oakland residents will be able to hike or motor the three miles to the park from the end of the Park Boulevard street car line.*
>
> *Undoubtedly one of the first actions taken with actual acquisition of the park will be preparation for establishing summer camps within the area where hundreds of Oakland people may spend their vacations.*
>
> *One may take his bedding on his back and hike from Oakland to a campsite under the redwoods.*[238]

Park department officials contemplated building golf courses and observation towers for visitors to gaze at San Francisco Bay. "An inn embowered by redwoods with facilities for dining, and probably many small cabins," as Louis Allen described it in the *Oakland Tribune*, was also part of the vision. Naturalist Henry A. Snow had collected a small menagerie of wild animals on a 1922 safari to southern Africa. These were to be placed in a zoo in the new park, as well.

Mayor John L. Davie expressed support but emphasized that a bond issue would be the preferred funding method rather than a tax levy. The city parks board placed a $538,000 bond issue on the August 29, 1922 ballot for the purchase of the 1,500-acre proposed "Sequoia Mountain Park." The voters rejected the bond issue.

Later, the city council voted on January 14, 1924, to purchase 180 acres—"Havens Redwood Grove on the slopes of Redwood Peak."[239] That plan did not move forward, either.

Snow's menagerie, known as the Oakland Municipal Zoo, was then being kept at Nineteenth Street and Harrison Streets.[240] Neighbors objected, maintaining that the wild animals were a nuisance and should be removed. Edgar M. Sanborn, vice-president of the park board, announced that the zoo would be relocated to Sequoia Mountain Park *if* the city passed a $20,000 line item in the 1925–26 budget. This was successful.

Over the next decade, park advocates shifted their emphasis to the burgeoning deal with EBMUD. By 1936, Oakland had assembled much of its own desired parkland. The Catholic archdiocese, in preparing to sell its Redwood Peak land to the new park district, also transferred its downslope acreage—with the "Big Trees" area—to the city. This land made up most of Sequoia Park. It extended from Joaquin Miller Park, which comprised Juanita Miller's home. The boundary between the Hights and Sequoia Park was near the lower intersection of Crane Way.

The City of Oakland renamed Thirteenth Avenue Park Boulevard in the 1920s and '30s when the redwood forest was reserved for parkland. The old logging road was the most direct route for city dwellers to access the open space from the Number 18 Key System streetcar.[241]

Snow's zoo was set up in the park, near the area where the Woodminster Amphitheater would later be built. Lloyd Graham Donaldson recalled riding his horse past the zoo in his boyhood. "It was scary! We could hear the lions roar when my mother and I would come down Joaquin Miller Road, which was an unpaved two-lane road then." The zoo was moved to its present location in Knowland Park in 1939.

Watershed Land Becomes Regional Parks

In the late nineteenth century, like mushrooms popping up after a rainstorm, private water companies proliferated throughout the East Bay hills. With their objective of protecting drinking water from pollution, the companies fenced in their open space lands, denying access to the public.

In the wake of a major drought in 1923, it was evident that the catchment basins and reservoirs built by Anthony Chabot were no longer adequate for the growing population. Voters approved a new water supply system. The pristine waters of the Mokulmne River were collected behind the Pardee Dam and piped to the East Bay. The East Bay Municipal Utility District was the system administrator. In 1929, the new utility district declared that ten thousand acres of scenic land was surplus and available for purchase.

Amid great interest by real estate developers and conservationists alike, one Berkeley hiker moved swiftly: Robert Sibley, executive manager of the University of California Alumni Association.[242] "The day it was reported in the newspapers that the EBMUD was going to give up its holdings here in the hills, he went right down to city officials and said, 'These valuable pieces of land ought to be preserved forever,'" recalled Carol Sibley.

Sibley brought out all the big guns. He immediately enlisted Hollis Thompson, Berkeley city manager, to organize the East Bay Regional Parks Association. Its ambitious goal was to obtain *all* the water company land for a chain of parks extending from Wildcat Canyon in the north to Lake Chabot in the south.

Harold French, with the Contra Costa Hills Club and the Sierra Club, sought the help of kindred civic organizations.

Samuel C. May, a professor of political science and director of the university bureau of public administration, had been lecturing about the need for parks for the previous decade. He met with interested civic groups to craft a unified approach.

May approached the owners of Kahn's Department store on Broadway in downtown Oakland. Their charitable foundation, with a mandate to serve public needs, had been formed in 1929 in accordance with the will of Frederick Kahn.[243] "Frederick Kahn, my uncle, was a great lover of the out-of-doors," Irving Kahn explained. "All during his lifetime, he had made it a habit to walk and hike through the hills and the beautiful open country just back of the East Bay cities. It was one of his greatest pleasures to hike over those hills and enjoy the magnificent and inspiring view of the growing East Bay cities, San Francisco Bay, and the Golden Gate."[244]

The Kahn foundation provided the significant sum of $5,000 to UC-Berkeley to sponsor an authoritative survey of local parklands. This study, published in 1930, was the work of a distinguished team: the Olmsted Brothers, architects; and Ansel F. Hall, representing the Educational Division of the National Park Service. The brothers were the sons and partners of Frederick Law Olmsted Sr., acclaimed designer of Central Park in New York City.

Ansel Hall had served as one of the first rangers in Yosemite and Sequoia National Parks. He quickly rose through the ranks to become chief forester and chief naturalist. From 1933 to 1938, Hall led the first scientific expeditions into Monument Valley in Arizona and today's national parks of Utah, thereby introducing them to the public.[245]

Enthusiasm for the parkland concept swept the East Bay. More than one thousand local citizens converged on the Hotel Oakland on January 29, 1931. Their purpose was to petition EBMUD to create a ten-thousand-acre park. Robert Gordon Sproul, University of California president, chaired the meeting.

But the utility, under the directorship of George C. Pardee, declined to manage the park. This was seen as too burdensome for an agency whose mandate was to provide water to a growing network of cities.

That left the park advocates with the task of creating a new agency. Never before had a special district been created for a park system. In order to succeed, support from elected officials was required. "The mayors of the East Bay cities were agreeable, and in March 1933 they organized a semi-official Regional Park Board under the chairmanship of Elbert M. Vail of Oakland," wrote Mimi Stein in *A Vision Achieved*.[246]

Assemblyman Frank K. Mott, a former mayor of Oakland, drafted AB 1114 to establish the new park district. After it was duly approved by the legislature, Governor James Rolph signed it into law in 1933.

Now the decision to create the new parks agency was up to the voters. Park advocates placed an initiative on the November 1934 ballot in Alameda County[247] to create the district and elect a board of directors. Significantly, the voters had to also levy a tax of five cents per each $100 of assessed real property valuation—during the depths of the Great Depression.

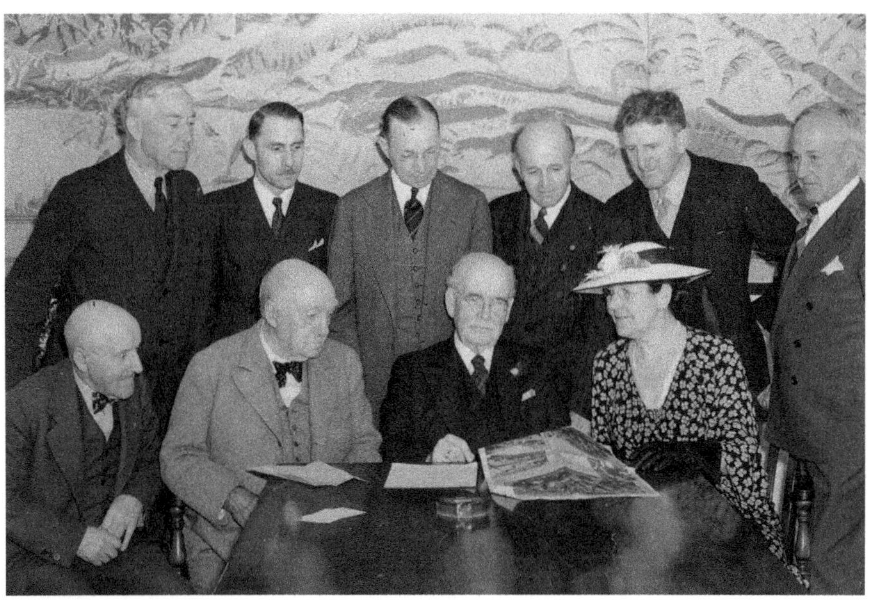

Founding directors of the East Bay Regional Park District meet with dignitaries in 1934. Standing, left to right: August Vollmer, Nil Aanonson (WPA director), Leroy Goodrich, Frank A. Kittridge (NPS director), Roy C. Smith (NPS inspector) and Elbert M. Vail (EBRPD first general manager). Seated, left to right: Thomas J. Roberts, John McLaren (creator of Golden Gate Park), Charles Lee Tilden and Aurelia Reinhardt. *Courtesy East Bay Regional Park District Archives.*

The Committee of One Thousand, led by Sproul, campaigned vigorously. The electorate approved the initiative with a remarkable 71 percent of the votes cast.[248]

Major Charles Lee Tilden of Alameda, a banker and veteran of the Spanish-American War, became the first chairman of the district board of directors. Mills College president Aurelia Reinhardt, labor union leader Thomas J. ("Tommy") Roberts, Berkeley police chief August Vollmer and Oakland attorney Leroy Goodrich were the other directors on that first board.

Elbert Vail was promptly appointed general manager. He hired Georgette Morton as office administrator. Both served without pay until the district financial system could be set up in 1936.

Tilden and Pardee Negotiate the First Land Transfer

Negotiations began in June 1936. The mission was to arrange the terms of a land purchase of ten thousand acres from EBMUD. While the park district and the water utility were the principals in the deal, other public agencies were affected as well. Mayors of seven local cities, plus Alameda County,[249] were interested in the outcome of the negotiations.

Representing the financial interests of the water district was George C. Pardee. Then in his late seventies, Pardee was a physician by profession. As California governor from 1903 to 1904, he had mounted a successful campaign to eradicate bubonic plague in San Francisco. A progressive Republican and ally of Theodore Roosevent, Pardee championed higher education, public health and environmental conservation. His name appears on rosters of park advocates.

Representing the new park agency was Major Charles Lee Tilden. Both because of his heroism in the Spanish-American War and his personal qualities, Tilden was held in high esteem by his fellow board members. Tilden, as board president, and Tommy Roberts, as secretary, led the EBRPD negotiators. Earl Warren, Alameda County district attorney, provided legal counsel.

Although he had a long history as an advocate for public access to open space, Pardee also had a fiduciary duty to his agency to collect the highest possible sale price.

The opening bid from Tilden was $1 million for the ten thousand acres. Pardee wanted $6 million. Tilden "stood unrelenting" against the EBMUD demand that the park district pass a bond issue to increase the offer price. Pardee wanted to receive a lump sum. Yet the only assets of the nascent park district were tax appropriations that would come in gradually. The annual park budget at the time was $194,000. Negotiations ground to a halt while an independent appraisal was obtained, at the request of the park district.

Meanwhile, Tilden had an opportunity to purchase sixty acres at the Redwood Bowl from rancher Alexander E. McNee. Appreciating the scenic value of this prime property and being aware that time was of the essence to secure it at a good price, Tilden advanced his own funds: $35 per acre, for a total of about $2000. Tilden was also aware that this sale price would hold down the appraised valuation of nearby EBMUD land.

The Relationship Between Tilden and Pardee

In telling the "origin story" of the park district, its official historians[250] have sought to explain the lengthy negotiations with an amusing anecdote told by Elbert Vail in his memoirs:

> *Both Pardee and Tilden were members of the first class to graduate from the University of California at Berkeley in 1878. The boys went to Mills College for their dates and dances. Tilden was the head of the U.C. band, while Pardee led the glee club.*
>
> *Tilden liked to tell about the time that he learned that the glee club had been invited to a dance at Mills but the band had not. To remedy the situation, Tilden went over the hill back of the university into Wildcat Canyon and hired the farmer's hay wagon and horse to take the band to Mills. It took over two hours....By the time they arrived, Pardee's glee club was serenading the girls. Cutting loose with their own drums and horns, the band drowned out the singers. Pandemonium resulted, but peace was finally restored with a compromise—the band played and the boys sang.*[251]

Jerry Kent, however, feels that this tale misrepresents the nature of the negotiations between Pardee and Tilden. Kent, who began his career with the East Bay Regional Park District in 1962 as an entry-level groundsman, rose through the ranks, eventually serving as assistant general manager of operations. Since his retirement in 2002, he has studied the history of the park district in great detail. "There's no way those negotiations were stalled because of an old incident between two college boys and a band," Kent maintained.

Tilden may have told that story later in life to illustrate that he and Pardeewere not cronies.

Both Tilden and Pardee were park advocates, in addition to both being very successful men in their late seventies. Tilden was one of the richest men in San Francisco.

This is not a story of two guys who had a college rivalry. This is a story of how two agencies, with possibly conflicting financial interests, worked together to craft a solution that would benefit the public by preserving parkland.

How did these two cagey old guys work together to make this land acquisition happen?

The early park advocates had three options, assuming that the "surplus land" was not sold for private development.

The first option would be for EBMUD to continue to own their "surplus" acreage and operate the land as a park. The second option would be to form a new park agency, funded with a bond issue. The third would be to form a new park agency, setting up a taxation scheme so that it would be "pay as you go."

The park advocates initially were perfectly okay with having EBMUD operating the parks. But this was not what Pardee wanted.

During the 1920s, when EBMUD needed to fund the new reservoir on the Mokelumne River, it was Roscoe Jones, a prominent attorney in Oakland, who ran the campaign that raised EBMUD $62 million dollars. When the park advocates formed committees, Jones was named chair of the committee in charge of implementing the Olmsted-Hall park plan. Jones was seen as the logical figure to play this role, as he was the ultimate insider with the water company. Later, when Pardee retired from the EBMUD board, Jones would be the person to replace him.

By 1931, within EBMUD, there was a difference of opinion as to how to fund the conversion of surplus watershed land to park land. Pardee was adamant that the water company should not go into the business of operating parks. Jones took the position there should be neither a bond issue nor a new park agency.

In January 1933, after two years of deadlock around who should own and operate the new parks, the nine local mayors from the two counties formed a committee to take action. Elbert Vail was appointed chair. He was the current chair of the Oakland Planning Commission, a businessman with a background in finance who had managed parks in three cities.

> *Vail obtained a letter from Olmsted and Hall recommending that a new park agency be formed. Within four months, Vail had initiated the legislation that led to the creation of an East Bay Municipal Park District.*
>
> *After the election, Pardee took the position that the park district should go to the voters for a three million dollar bond issue. In return, EBMUD would give the park district the watershed land and cut the tax rate it was charging the public.*
>
> *The mayors objected. Their position was that during the depths of the Great Depression the public would not grasp that EBMUD was essentially offering a no-cost deal. They wanted to see a pay-as-you-go arrangement for the park district to pay the water company. But the mayors conceded that this would only "pencil out" if two thousand acres were transferred, not the entire ten thousand acres the utility had available. The rest of the land title could be transferred later.*

One aspect of the story that is not well known is that the federal Works Progress Administration could bring pressure to bear. The leadership of the WPA and the local mayors all wanted the contracts to be finalized. They wanted to put their thousands of men to work.[252] Once the first land had been put in the name of the park district, the effort would become eligible for federal funding.

Tommy Roberts contacted the mayors of affected cities to ask for their support. He advised them that WPA money was available to develop the parks and that job opportunities would be available for local residents.[253]

Similarly, Tilden, Goodrich and Roberts told the Alameda County supervisors that the time was ripe to adopt a pay-as-you-go method in order to engage with the WPA. The park district would levy a tax of five cents per $1,000 of assessed property value, while the water company would cut its levy by five cents, thus making the deal "revenue neutral" to the public.[254]

The compromise was finally crafted in June 1936. Earl Warren, representing the park district in his capacity as Alameda County district attorney, is credited with having formulated the final terms.[255] The park district purchased 2,166 acres—comprising today's Temescal Recreation Area, Tilden and Sibley Regional Parks, as well as part of Wildcat Canyon. Redwood Regional Park would be acquired later.

The price was $656,544, or just over $300 per acre. Over the course of the next five years, the funds would pass from tax revenues paid to the park district directly to EBMUD. As the funds were received, the water utility cut its tax rate, so the public would pay the levy only once.[256]

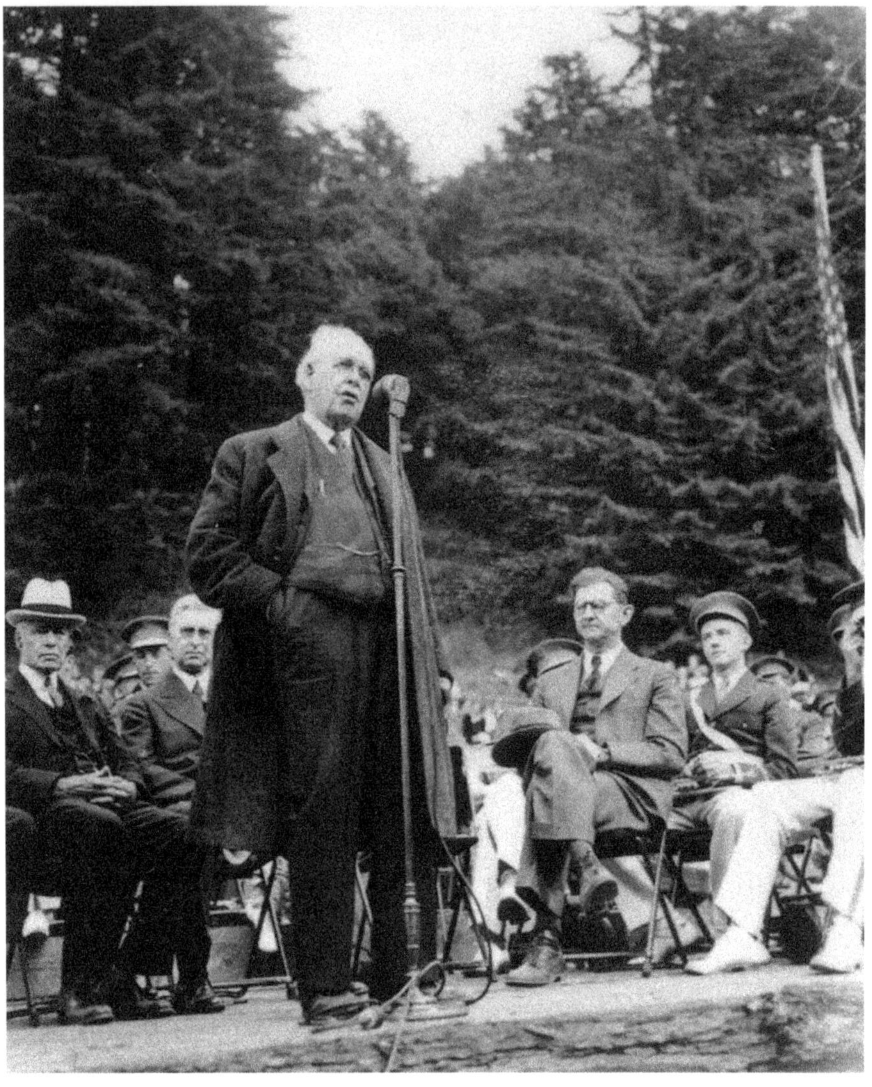

Major Charles Lee Tilden addressed the audience at that opening ceremonies for the park district, held at the Redwood Bowl. *Courtesy East Bay Regional Park District archives.*

Tilden, meanwhile, had been wise to purchase the McNee property at the Redwood Bowl for $35 per acre. Prices for subsequent land purchases were more than ten times higher.

The grand opening ceremonies for the park district were held at the Redwood Bowl on October 18, 1936—even before the district actually owned Redwood Regional Park. The University of California marching band performed. Equestrians from Mills College and western-style riding clubs paraded. Local archers and fencers gave demonstrations. Major Tilden gave the keynote speech.

New Deal Agencies Build Park Infrastructure

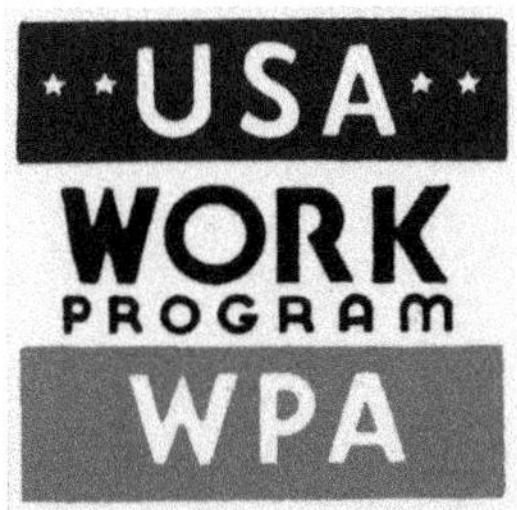

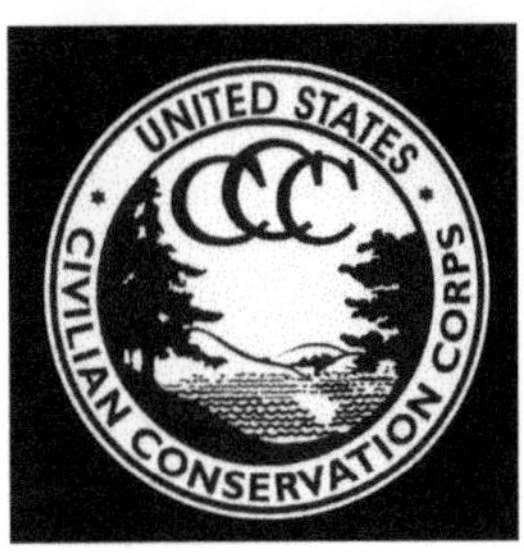

Top: Crews from the Works Progress Administration (WPA) built the Woodminster Amphitheater, as well as roads, bridges and other park infrastructure. *Courtesy East Bay Regional Park District archives.*

Bottom: Crews from the Civilian Conservation Corps (CCC) felled eucalyptus trees, cleared the Redwood Bowl and built the Stream Trail picnic areas. *Courtesy East Bay Regional Park District archives.*

An enormous amount of work was needed before the public could be invited into the new parks. But after the initial land purchase was made, federal assistance brought in the labor armies of the New Deal. Shovel-ready projects beckoned to American men who were eager to work.

One of the first jobs completed was the construction of Skyline Boulevard, along the crest of the Oakland hills. For this, unemployed local residents were hired through the State Emergency Relief Administration.[257]

Elbert Vail and Tommy Roberts Take the Reins

Elbert Vail soon proved his expertise as a project manager. Networking widely, he procured $3 million worth of labor and materials.

Tommy Roberts, with his background as a union negotiator, was elected secretary of the EBRPD board of directors. He facilitated intra-agency arrangements.

Vail outlined the work that would be needed in order to develop parks. Much of the New Deal manpower would be concentrated at Tilden Park, where Camp Wildcat Canyon was the largest base of operations. A combined park headquarters building and beachhouse was designed for the shore of Lake Temescal.

The CCC in the East Bay Redwood Region

The Civilian Conservation Corps (CCC) was a program for unskilled laborers, young men between the ages of seventeen and twenty-eight, from impoverished communities throughout the country. They were provided with room and board and a small stipend of thirty dollars per month, twenty-five of which was sent home to the enrollees' families.

Author Michael Hiltzik has written extensively about the New Deal work programs:

> *The average CCC enrollee stayed for nine months, during which time he gained up to thirty pounds, thanks to three meals a day served up by the army quartermaster as fuel for eight hours of daily labor. For many boys, this was more abundant fare than they had ever seen, a full stomach being as novel an experience as the indoor plumbing and electricity of the CCC barracks.*[258]

CCC Camps

The Civilian Conservation Corps housed its workers at two camps in the redwood region of the East Bay hills.

Camp Grass Valley was located on a ridge east of the Oakland Lake Chabot golf course. From this vantage point, workers could enjoy views over San Francisco Bay when fog conditions permitted. The site remains largely unchanged eighty years later. A row of eucalyptus trees marks the camp boundary, and the access road once used by heavy trucks is now the popular Goldenrod Trail.

Camp San Leandro Reservoir was originally intended to be a segregated camp for African American enrollees. In the era before the United States armed forces were integrated during World War II, and the civil rights movement of the 1960s followed, racial segregation went largely unquestioned.

Leading CCC, however, was Roosevelt cabinet secretary Harold Ickes, a white man who was an early advocate for racial equality. Under his administration of the CCC, enrollees benefited from a recruitment policy barring discrimination "on account of race, color, or creed." The frugal Ickes did not allow power tools to be purchased. He wanted to reserve more money for salaries to put more men to work.[259]

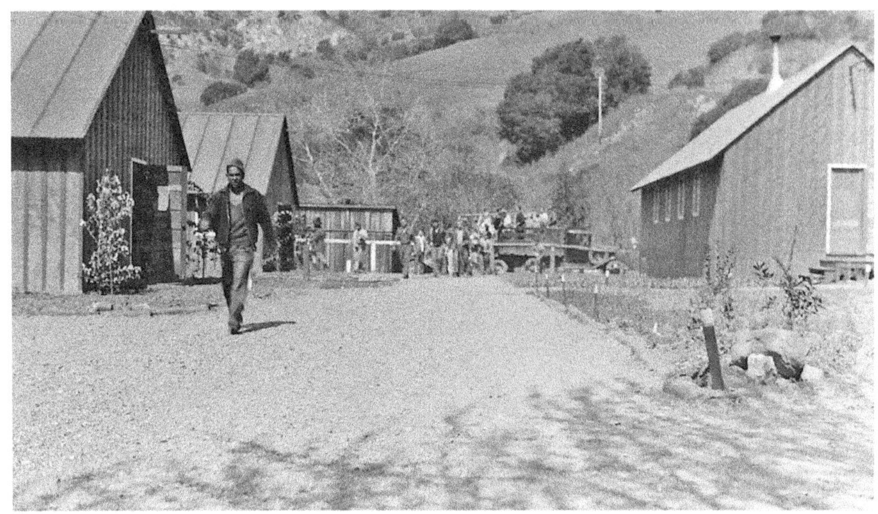

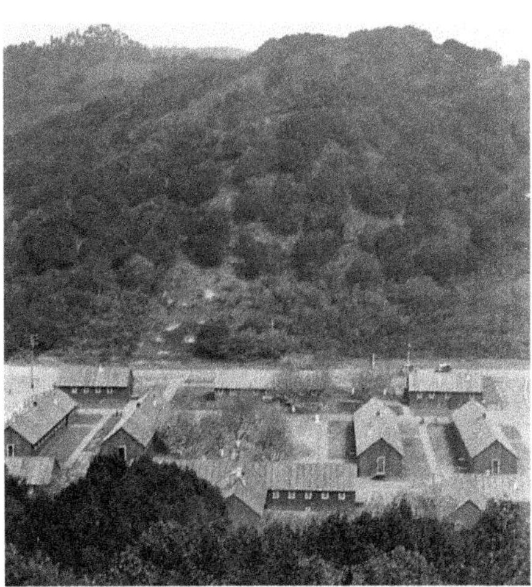

Above: A CCC enrollee hurries to the mess tent. This was "chow time" at Camp San Leandro Lake. *Courtesy East Bay Regional Park District archives.*

Left: CCC Camp San Leandro Lake was originally a segregated site housing African American men. It was located by today's Lake Chabot. *Courtesy East Bay Regional Park District archives.*

On the other hand, Roosevelt cabinet secretary and Works Progress Administration (WPA) administrator Harry Hopkins favored high-profile projects with symbolic value. Under his leadership, WPA stonemasons—many of whom were skilled Italian immigrants who lived in the Temescal district—built bridges, walls, culverts, restrooms, hiking structures and drinking fountains throughout the parks.

The CCC crews worked on the golf course itself, as well as in Grass Valley, Redwood Park and beyond. WPA workers, who were both male and female and of any age, typically remained at home with their families.

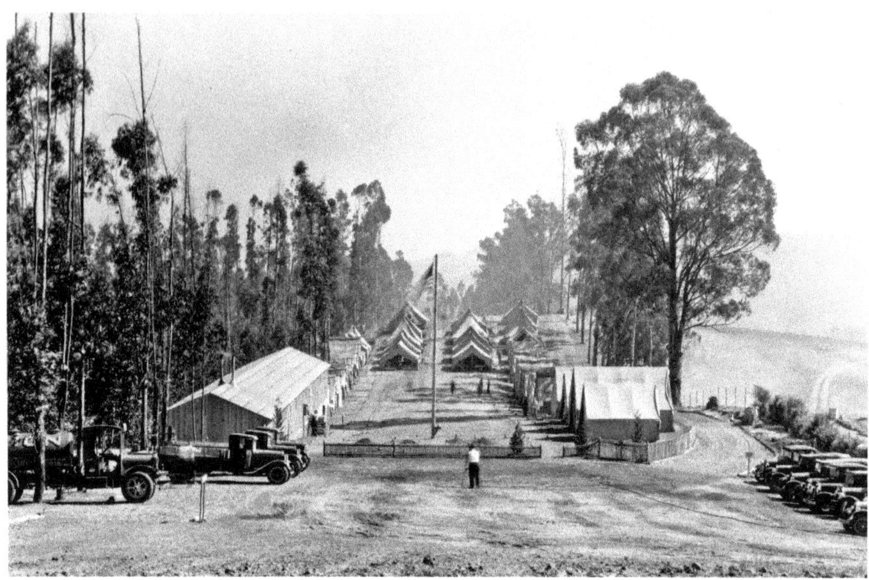

CCC crews staged from Camp Grass Valley, located on a ridge above the city of Oakland golf course, along today's Goldenrod Trail in Anthony Chabot Regional Park. *Courtesy East Bay Regional Park District archives.*

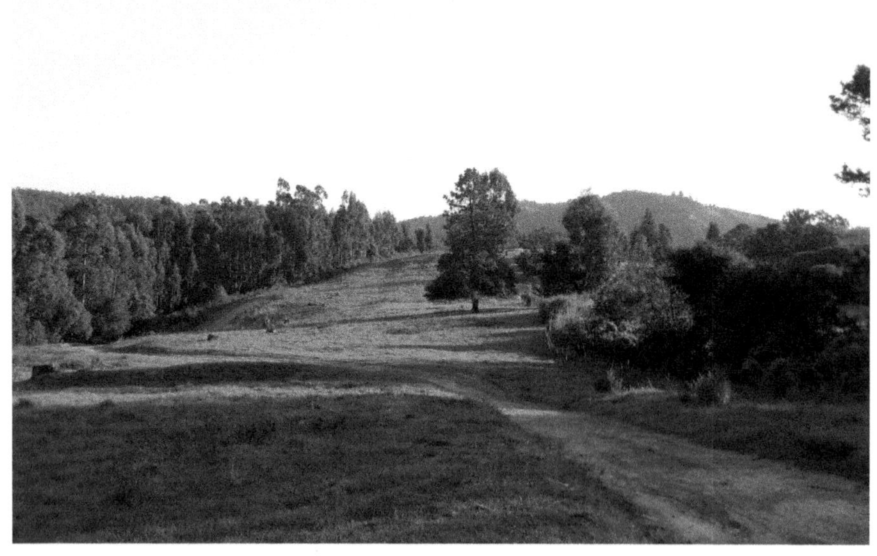

The site of Camp Grass Valley remains open space on the Goldenrod Trail in Anthony Chabot Regional Park, upslope from the Oakland Municipal Golf Course. *Amelia S. Marshall photograph.*

CCC laborers did the hard work of clearing fields. Director Tommy Roberts could often be found at the worksites with the young laborers. "Rain or shine, Director Tommy Roberts walked among the WPA workers offering them encouragement," Elbert Vail wrote in his unpublished history. "After the roads had been built, we began construction of caretakers' residences, office buildings, campgrounds, and restrooms."[260]

In Grass Valley, now Anthony Chabot Regional Park, CCC workers were the first crews in. They freed meadows and hillsides from the tangle of eucalyptus trees generated by the Havens tree plantations. Poison oak was cleared as well, resulting in significant down time for the projects and misery for the workers.

The CCC crews cleared the Redwood Bowl, making room for the regional parks grand opening ceremony on October 18, 1936. These open fields provided space for the archery range and cross-country course that would be operated for a century to come by the Redwood Bowmen club.

The CCC also opened up fields in the Stream Trail area of Redwood Regional Park, including the old mill sites and the pioneer orchard area.

The WPA in Redwood Regional Park

For the first three decades after the inception of the EBRPD, Redwood Regional Park included the area that is now Roberts Regional Recreation Area. Archival documents such as Elbert Vail's annual reports on WPA work do not often provide specific locations where projects were done. While a few specific structures—such as the Little Church of the Wildwood—are known to have been constructed by New Deal workers, precise identification of all New Deal projects remains elusive.

WPA crews built the stone bridge where Redwood Road crosses the creek. Within the stream trail area, the WPA bridged the creek where sawmills had stood seventy years earlier.

In the towering Aurelia Reinhardt redwood grove, the WPA rebuilt the rustic Church of the Wildwood. Originally an exhibit at the 1939 World's Fair on Treasure Island, the church was a popular wedding chapel until it burned down around 1958.

Lloyd Graham Donaldson recalled riding his horse under WPA-constructed camping shelters that were built at the Fern Dell and Trail's End picnic areas. District managers concluded that the original stone structures

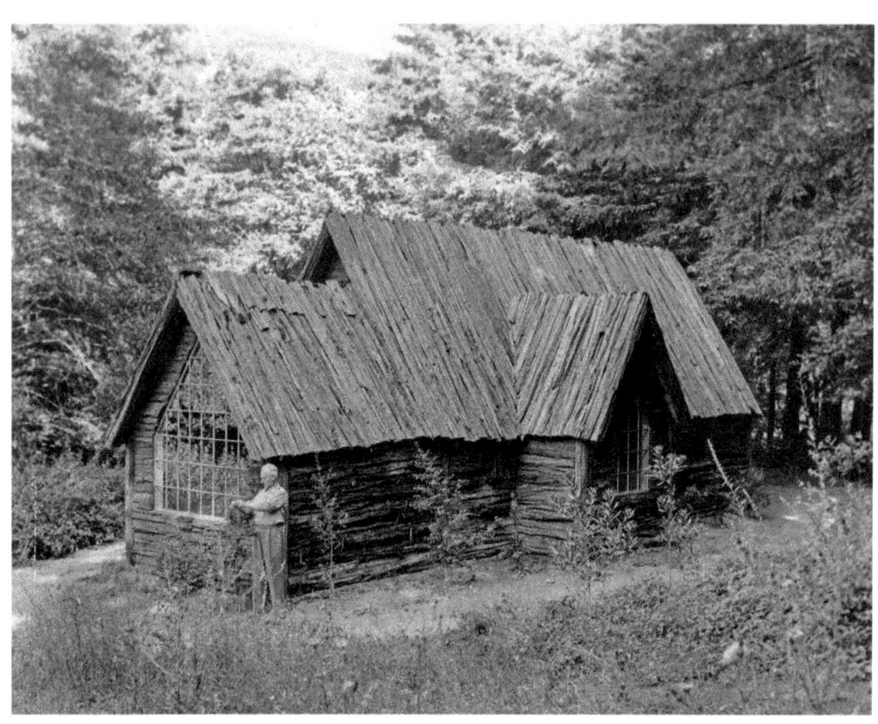

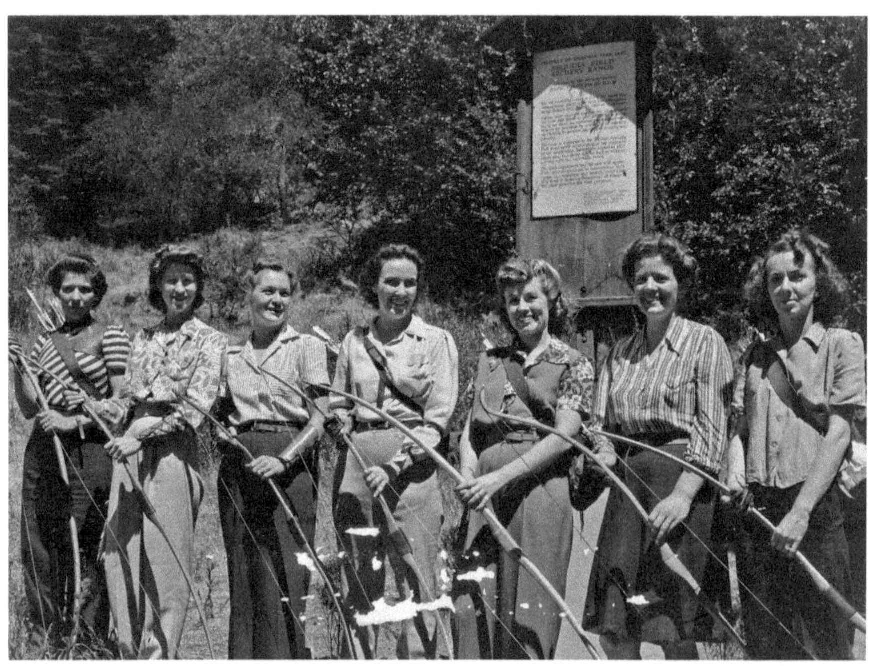

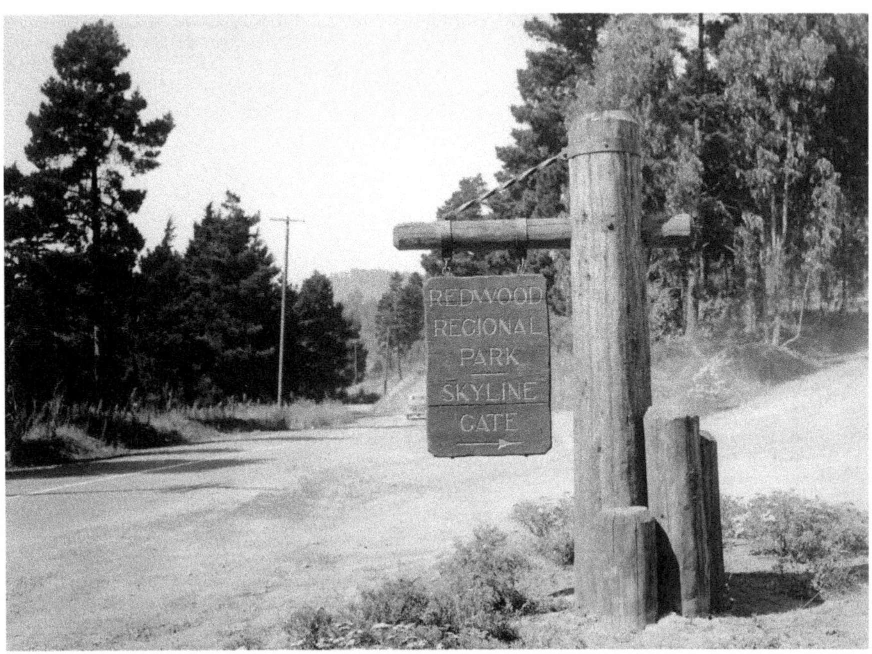

Above: Skyline Gate to Redwood Regional Park, site of the WPA-built "warden's cottage," was called "Five Points" by local equestrians. *Courtesy East Bay Regional Park District archives.*

Opposite, top: The Little Church of the Wildwood was moved from Treasure Island in 1939 by the WPA and reconstructed along Redwood Creek. Elbert Vail, first park district general manager, stands by the church, circa 1940. *Courtesy East Bay Regional Park District archives.*

Opposite, bottom: Archers enjoy the range at the Redwood Bowl. The Redwood Bowmen have maintained the archery range since approximately 1939. *Courtesy East Bay Regional Park District archives.*

would be hazardous in case of an earthquake, so park staff subsequently rebuilt them using modern methods.

The report written by Elbert Vail in 1940, in which the New Deal work is described, mentions the construction of a "warden's cottage," the park residence located at the Skyline Gate (or "Five Points") staging area.

The WPA may have also done work on the wood and stone park residence in today's Roberts Recreation Area, where the Howland family cared for the elderly Tommy Roberts in the 1940s. One of the last WPA projects done in the area was paving Redwood Road. The distinctive stone bridge over Redwood Road at the park entrance is believed to have been WPA work.

The WPA in Grass Valley/Anthony Chabot Regional Park

One of the most enduring examples of WPA work in Anthony Chabot Regional Park is the Stone Bridge, which connects the (Claud) Brandon Trail and Grass Valley Trail.

While it is tempting to imagine that the WPA built the rustic stone drinking fountains along the Goldenrod Trail, south of the Clyde Woolridge staging area and north of the site of Camp Grass Valley, their origins are unproven. While these drinking fountains are of similar construction to known WPA work in Tilden Park, archival photographs from the Metropolitan Horsemen's Association show equestrians unveiling similar fountains, with side-by-side troughs for people and horses, in the 1950s.

The WPA in Oakland City Parks

In Joaquin Miller Park, the Woodminster Amphitheater remains the most beloved monument to WPA craftsmanship. The stone cascades, Art Deco bas reliefs and concrete stage are a lasting reminder of the New Deal legacy.

Less beloved are the tons of concrete left by New Deal work crews in Sausal Creek Canyon. Before modern ecological awareness influenced urban creek management, "channelization" was seen as an appropriate strategy to prevent flooding downstream.

Consequently, hikers who search along the Sausal Creek Trail for relics of Caspar Hopkins's dam or pipeline will encounter only pipes, rebar and concrete from the New Deal era.

The WPA Beyond the Redwoods

While Sibley Volcanic Regional Park, originally known as "Roundtop," was one of the five original East Bay Regional parks, it contains no known evidence of WPA or CCC work. Residences in Sibley Park were constructed subsequent to the New Deal period by park staff, according to Jerry Kent.

The WPA made major contributions in the Lake Temescal park, however; in addition to capital work on the dam and reservoir, the WPA constructed the Temescal Beach House, which served as the first park district headquarters.

In addition, the WPA filled the area behind the dam to create playing fields and parking lots and built the artificial beaches on the east side of the lake, using many truckloads of imported sand. Stone benches and drinking fountains near the beachhouse are similar in construction to those in Tilden and Alvarado Parks.

Chapter 8

RANGERS

All Quiet in the Parklands During World War II

With the outbreak of hostilities, the park district administration entered a quiescent period. General Manager Elbert Vail retired in 1942 and was replaced by Harold L. Curtiss. The five-park district workforce decreased from thirty to only six, as men enlisted or took defense industry jobs. Managers discussed hiring women, but no one followed up on this idea.[261]

Wesley and Loraine Adams remained stateside to serve as supervisors of Redwood Regional Park. She took reservations for the hunting lodge near the park entrance and helped Wes with fighting fires, as needed. When fires broke out, Wes and Loraine Adams were assisted by civilian volunteers—the bartenders from the taverns on Redwood Road.[262] "Adams' original fire truck was a 1932 Studebaker touring car, painted red, with no bell, lights, or siren. In the back seat was a 50-gallon drum of water with a belt-driven pump and a garden hose."

Military Training in Redwood Regional Park

By 1943, Jim Howland and his family were living at the park residence near the Redwood Bowl. Howland, who went to work for the district in 1962, would later be supervisor of Roberts and Don Castro Parks. He told park

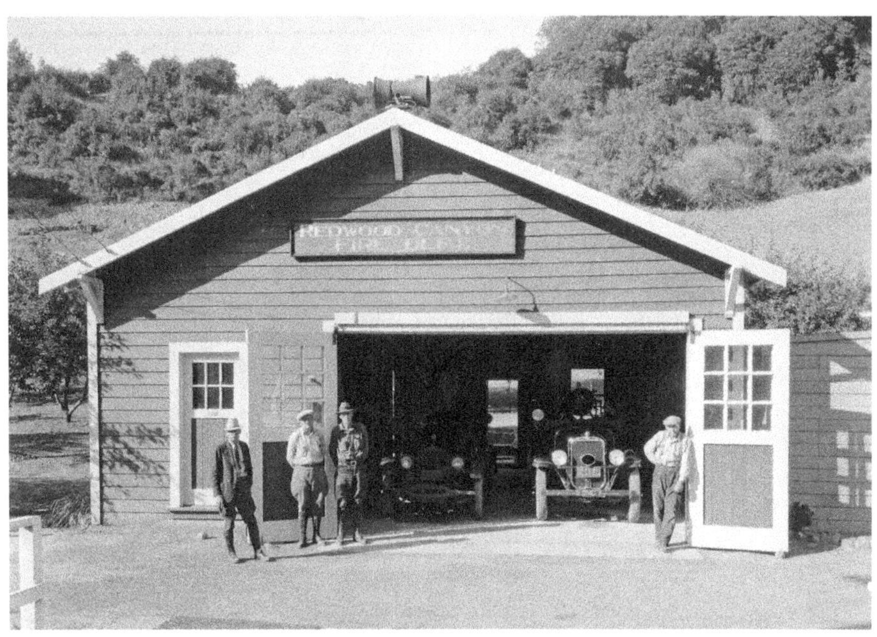

Redwood Canyon had established a fire protection district by 1931. The fire station has long been in place by the main entrance to Redwood Park. *Courtesy East Bay Regional Park District archives.*

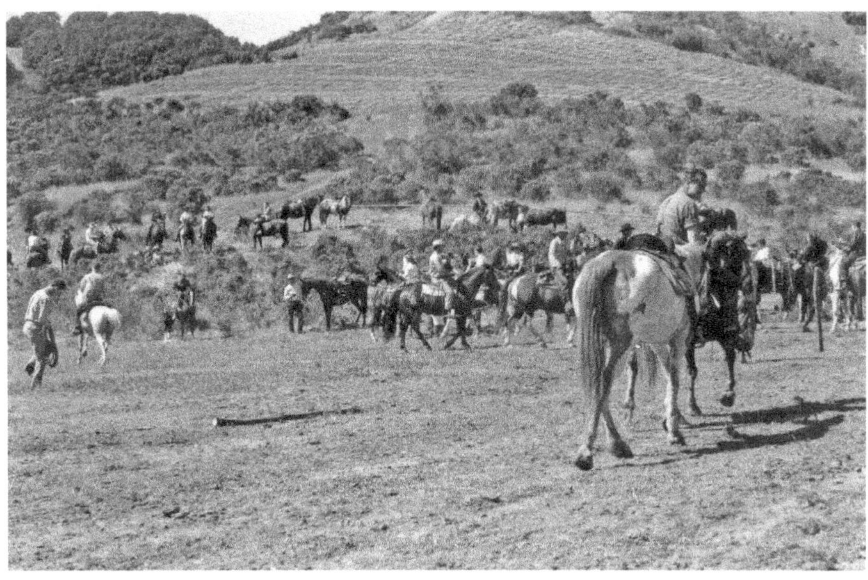

The Metropolitan Horsemen's Association held a field day in Redwood Park on May 20, 1945. In the background is a "practice graveyard" constructed by African American military troops. *Courtesy East Bay Regional Park District archives.*

historian Mimi Stein that his grandfather had gone to work for the park district in 1939, and his father two years later.

Howland recalled that an antiaircraft battery was stationed near today's Trudeau Training Center at 11500 Skyline Boulevard. An African American training battalion camped near the East Ridge Trail. They were assigned to dig "practice" cemeteries in the Stream Trail area, as Howland recalled:

> *There was a two-week survivor training program which pre-flight students went through. At the end of the course, they were given one basic C ration unit per day and sent out to try to survive on whatever they could find at Redwood. I remember they would always wind up at the Redwood Bowl. The instructor would always make sure that my mother had a pound of coffee on hand—it was hard to come by then—so she would be sure to have a hot pot of coffee ready for those guys after they'd struggled through two weeks of survival training.*
>
> *And what I loved about having them there was that most of these guys would save their little chocolate bars and C rations for me. All I saw of chocolate through the war years was what I got from them.*[263]

Acquiring Land for Redwood Regional Park

The Initial Land Purchases

The land that would become Redwood Regional Park and Roberts Recreation Area went through several phases of ownership. After the departure of the Native people, the ridgeline was the designated boundary separating the Antonio Peralta land grant from that of Joaquín Moraga. In the mid-1800s, American courts determined the area to be "Tract Number Two of Rancho Laguna de los Palos Colorados"—the Middle Redwoods. Early records of ownership changes are incomplete and sometimes mysterious.

During the latter decades of the nineteenth century, the United States government made a practice of giving public land to the railroads as an incentive to westward expansion. Using a strategy called "checkerboarding," the railroads were granted alternating USGS mapping sections, resulting in a pattern of public and private ownership that resembled a checkerboard. Part of the East Bay redwood lands evidently passed through temporary railroad ownership.

History of the "Archbishop's Parcel"

Thanks to the Roman Catholic Archdiocese of San Francisco, the provenance of one particular parcel in today's Roberts Recreation Area is well documented.

The "Archbishop Hanna Property" was the fifty-eight-acre parcel encompassing Redwood Peak and part of the Redwood Bowl. The boundary line between Contra Costa and Alameda Counties formed the western edge of the Archbishop's Property.

Land title records show that the archbishop's parcel was in the public domain prior to 1862, the year it was patented by the federal government to the Western Pacific Railroad through an Act of Congress. At some point, the railroad deeded it to Charles McLaughlin, a railroad contractor who died in 1883. His heirs then deeded the property to J.B. McNally for five dollars on September 20, 1895.[264] McNally was identified as the "pastor of St. Patrick's Roman Catholic Church," likely the same parish that is still located at Tenth and Peralta Streets in West Oakland.

In an 1896 article in the *San Francisco Call* newspaper promoting homesteading in the East Bay hills, area residents are mentioned: "Father McNally of Oakland has a ranch overlooking both counties, where he is considering the idea of colonizing some of the City's homeless boys and giving them a chance at useful lives. Near neighbors to him are J.T. Classen and C.C. Crowley, who will shortly build him a home overlooking San Francisco, Oakland, Alameda and the Golden Gate."[265]

McNally deeded the tract to "the Roman Catholic Archbishop of San Francisco, a corporation sole, in trust" on September 11, 1911. In return, he wanted the proceeds from the sale of the property used for the St. Patrick's Parish School for Girls, run by the sisters of St. Joseph, and the School for Boys, run by the Christian Brothers, as well as to maintain the schools "in a state of efficiency."

At the time Father McNally handed his property over to the archdiocese, the man in charge was San Francisco Catholic archbishop Patrick W. Riordan (1841–1914). Later, when the property was sold to the park district, the archbishop was Edward J. Hanna (1860–1944). Hence, park district maps cite the "Archbishop Hanna Property."

Within the Catholic archdiocese, both Riordan and Hanna presided over periods of real estate acquisition and construction. Riordan built many churches and seminaries. After comforting victims of the 1906 San Francisco earthquake, Riordan announced, "We shall rebuild!"

Hanna also acquired property for the church. He moved Catholic cemeteries from San Francisco, south to the more spacious Colma. He was a close friend of San Francisco mayor, and later California governor, James ("Sunny Jim") Rolph. The archdiocese also owned a significant amount of acreage in what is now Oakland's Joaquin Miller Park, according to EBRPD land transfer documents.

Tommy Roberts, negotiating with the archdiocese on behalf of the EBRPD, was evidently not a pushover:

> [R]*egarding the East Bay Regional Park District's interest in securing the McNally property, we have already offered what we consider even more than the property is worth...*
>
> *Whereas our top offer is $500 an acre for the McNally property, we have recently purchased 100 acres, known as the Cornelius Property with frontage on both Skyline Boulevard and Redwood Road at $250 per acre...*
>
> *We have no desire to take advantage of the Church in any way, or to drive a hard bargain. However, we do feel that with this property being in our hands, for no commercial purpose, but only in the interest of giving the people of the Bay Area healthful recreation, our offer is entirely fair.*
>
> *We understand that the McNally Property has been held on the tax books of both Alameda and Contra Costa Counties at an extremely low price. In Alameda County its assessed valuation is $8.00 per acre. The land is cut off from all utilities, has no improvements, and is serving no worthwhile purpose.*[266]

According to Lloyd Graham Donaldson, the church had at one point intended to build a monastery at the site. By the mid-twentieth century, however, urban development was encroaching. The site had become unsuitable for an isolated religious community.

The J. T. Classen Family Farm and Orchard

Far less is known about J.T. Classen, Father McNally's neighbor, whose family had a farm and orchard near where today's Roberts Park staff residence and swim center are located.

One of the most intriguing tales of the Classen family appears in the September 7, 1941 edition of "The Knave" local history column in the *Oakland Tribune*.[267] John S. Engs, an Oakland dentist, reminisced about his Sunday horseback rides to Redwood Peak in the early years of the twentieth century:

> [The Classens lived] *in a house on the crest of the ridge at the base of Redwood Peak. He was the owner of a rather large tract of land adjoining. On this tract grew many second-growth redwood trees. Back of his house was a fruit orchard which had been planted in a deep depression in the hilltop. He told me that in the early days, the area covered by his orchard had been a mill pond.*[268] *The water had been impounded by a dam at the lower end. Many years ago, the dam broke, releasing the water. He said "I have many fine second growth trees on my place, and some day they will be worth money."*
>
> *For the sum of twenty-five cents Mrs. Closson* [sic] *served a bountiful luncheon of bread and butter, ham and eggs, and coffee. For those bringing their own food she would furnish hot tea or coffee. I used to ride to their place nearly every Sunday morning for several years following the* [1906] *earthquake.*[269]

Additional information on the Classen family appears in the 1977 EBRPD *Redwood Regional Park Resource Study*: "Classen served homemade root beer and kept a guest register for the many people attracted to the area."[270] Local lore also holds that Classen had a fruit stand; he sold fruit drinks to hikers.

The story told by Dr. John Engs casts further light on the history of the property, which is now presumably near the Redwood Bowmen archery range in Roberts Park: "A few years later I head that a group composed of his regular Sunday visitors, with Judge Snook at the head, organized an outing club and purchased the property, including the house. I think their intention was to develop the place as a sort of clubhouse. Unfortunately, fire destroyed the house soon after."[271]

Park district land acquisition records[272] show that a parcel owned by Capwell and Snook was located just north of Redwood Peak, encompassing part of the Redwood Bowl, the tallest trees, and today's Redwood Bowmen archery range. Charles E. and Jennie Wade Snook sold the property to EBMUD in 1937. It was acquired by the park district a few years later. Could the Classen home have been at the site of the present park district staff residence near the Redwood Bowl in Roberts Park?

Paul Miller, who served as Roberts Recreation Area supervisor from 1996 through 2008, suspects that there has been a house at the park residence site since the nineteenth century. Local lore holds that the original house had burned down and been rebuilt. "I went down into the basement to inspect the old stone foundation and saw what looked like scorch marks on the inside walls," Miller noted.

The park residence has been the home to many staff people over the years. Notably, the Howland family lived there while caring for the elderly Tommy Roberts. Upon his passing, the park was named for the beloved park advocate and labor leader.

During his time as park supervisor, Miller encountered a park visitor who told him that he had lived there as a boy. According to the visitor, when he was twelve years old, he and his parents had come in 1942 from Missouri to Oakland for jobs in the Kaiser shipyards. The family rented a ranch there, with a horse barn and orchard, as Miller recalled:

> *He told me that their house was at the site of the present-day house next to the park office. The barn was at the site of the present-day bath house and swimming pool. There was a significant expanse of orchards from the Redwood Bowl through Roberts Park. There was a single surviving apple tree by the Redwood Bowmen's storage building while I worked in Roberts Park. Mature olive trees still grow on the west side of Redwood Peak.*

Another curious, but unsubstantiated, story told to Paul Miller dates from the 1850s:

> *A man named Jim Bissell who owned a tree company told me that the old bullwhackers used to water their stock at the lower end of the Redwood Bowl where the West Ridge Trail meets the Graham Trail. That means they had some source of water—either a spring or a pond for impounding rain water. The orchards, too, had some source of water that is not evident today.*

The Classen orchards once stood where today are the parking lots for Roberts Recreation Area. Photos of these orchards can be seen on Google Earth, directing the "history" feature at the west side of Redwood Peak.

Other Landowners in Redwood Regional Park

With the return of peacetime in the late 1940s, the park district undertook a vigorous program of land acquisition. Filling in the gaps to complete Redwood Regional Park was a priority.

One of the early landowners of the area along today's West Ridge Trail was George May, a successful wheat farmer and cattle rancher who donated a school in Livermore.[273]

Around 1900, grazing land was acquired by the McNee Company, an outfit that would become the largest private landowner in the East Bay redwoods.

Duncan McNee (1849–1913) is often described as "an early California land baron." His ranch in the northern Santa Cruz Mountains above Montara has been a state park since 1984. Alexander McNee (1878–1938), son of Duncan, owned hundreds of acres in today's Redwood and Roberts Parks and near the town of Canyon.

Alexander McNee became notorious only late in life. As a sixty-year-old bachelor, he met Florence Farnum Otis, proprietor of a gift shop at 5531 Telegraph Avenue, North Oakland. Two months later, McNee became her third husband. He died two months after the wedding. Numerous heirs contested the will, alleging undue influence. The litigants included Stanford University, which would have inherited one-third of his $100,000 estate had the widow predeceased McNee.[274] No wonder Major Tilden was able to buy the McNee parcel near the Redwood Bowl for $35 per acre a few years earlier!

Subsequently, the McNee Company extracted a higher price: seventy-five dollars per acre for two 160-acre parcels. The first sale, in 1944, was for 160 acres ("including a modern 65-stall horse stable,"[275] Piedmont Stables, which had been leased to the Piedmont Trails Club). The land was northeast of Redwood Road, upslope to the West Ridge. The second 160-acre parcel, sold in 1948, was in the southeastern Stream Trail area extending to the East Ridge. The McNee Company had also acquired the 160-acre Grimes Ranch in the upper end of the Stream Trail Canyon.[276]

The second-largest landowner was Thomas Bridge, who owned 365 acres east of the Capwell-Snook property, extending through the Stream Trail canyon and west to the East Ridge.This sale was accomplished in the mid-1930s.

EBMUD sold the park district 1,057.3 acres from its watershed corridor above Upper San Leandro Reservoir for $246,227 in 1943. EBMUD had bought some land in 1937 from Lila Havens (as Villa Site and Development Company) and also from Charles E. and Jennie Wade Snook (Central Development Company). The district purchased 83 acres in 1951 from a seller called "Bechdolt, formerly Brougher."[277] This would appear to have been the orchard area that is now the parking lots for Roberts Park.

Smaller parcels were sold by owners named White, Mollie Carey, Walling and Kish. Carey, Kish and Krum had acreage in the old sawmill area.

POSTWAR LAND ACQUISITIONS BY THE PARK DISTRICT

The park district bought the forty-acre Montiero Ranch in 1948 for $9,000. At today's Serpentine Prairie area, the district purchased the one-hundred-acre parcel of Albert Cornelius Jr. and Alice Highby. The property had "frontage on both Skyline Boulevard and Redwood Road." The sale price was $250 per acre in 1949. The property had previously been owned by Nellie Leise. This would appear to have been an L-shaped parcel, with Miss Graham's Redwood Riding Academy at the corner of Skyline Boulevard (then still called Joaquin Miller Road) and Redwood Road.

As property tax revenue accumulated, EBMUD relinquished an additional 92 acres, including the Redwood Park entrance, in 1951. "It took years to get EBMUD to sell the acreage that became RRP to the park district," Jerry Kent commented. In 1957, John S. and Blanche Desmond sold the district 0.19 acre at the Skyline Gate for $1,500.

Throughout the park district history, managers have had a policy to purchase any of the still privately owned real estate along Redwood Road that comes on the market. Beginning in the 1950s, the park district purchased the sloping hillside above Skyline Ranch from Mayer.

Sam Sparks was one of the major developers of the nearby Crestmont subdivision on the old Stauffer mining area. His Great Day Development Company sold the park district the hillside above Redwood Road, northeast of Skyline Ranch. His name lives on at Sparky's hamburger stand at the Lincoln Square shopping center.

RICHARD WALPOLE BRINGS RECREATIONAL ATTRACTIONS

Richard Walpole became the third park district general manager in 1945. A tall, imposing figure, he was an avid golfer. During a 1945 visit to Griffith Park in Los Angeles, Walpole was impressed with the recreational and revenue-enhancing possibilities of concessions in the East Bay regional parks. The Walpole era of the late 1940s and early 1950s saw the addition of the elegant Hershell Spillman merry-go-round, pony rides and golf course in Tilden Park.

In 1953, Walpole split off the old Classen orchard area was from Redwood Park. The new "recreation area"[278] was named after Tommy Roberts.

The grand opening of Roberts Regional in 1953 featured a new merry-go-round, pony rides, a ball field and a swimming pool, in addition to hiking

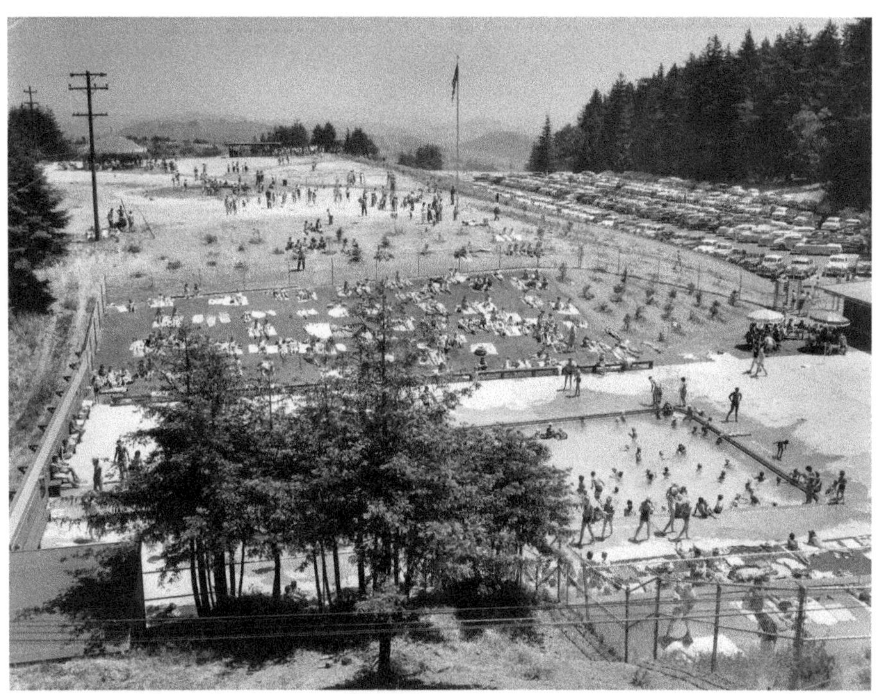

Crowds thronged to the new concessions at Roberts Recreation Area on opening day in 1953. *Courtesy East Bay Regional Park District archives.*

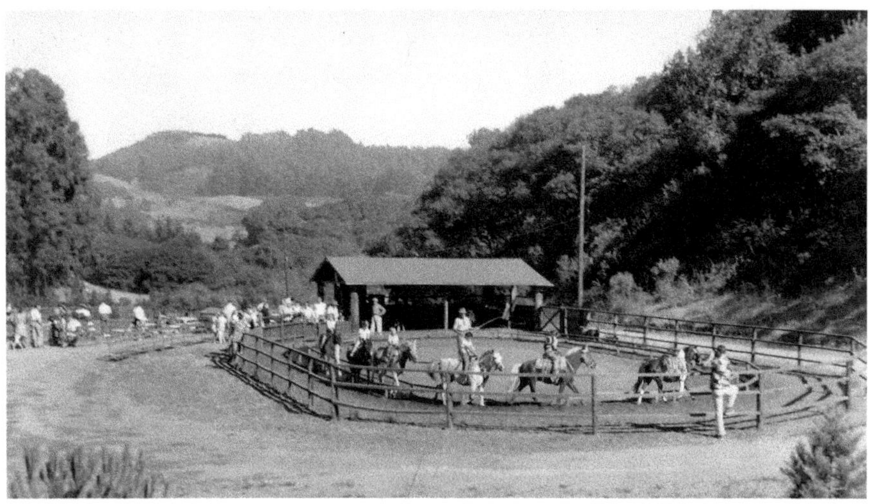

Little children could ride ponies at Roberts Park for about ten years, until the area was redesigned by William Penn Mott in the early 1960s. *Courtesy East Bay Regional Park District archives.*

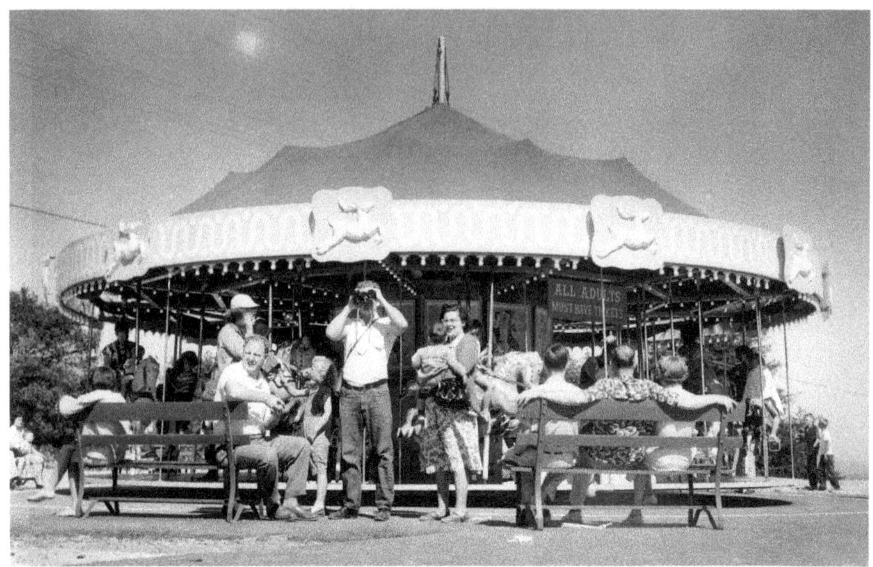

The merry-go-round was one of the recreation amenities that Richard Walpole brought to Roberts Recreation Area, inspired by his visit to Griffith Park in Los Angeles. Twelve years later, William Penn Mott directed that the concessions be closed down, to create more of a wilderness experience for visitors. The merry-go-round was moved to Kennedy Park in Hayward. *Courtesy East Bay Regional Park District archives.*

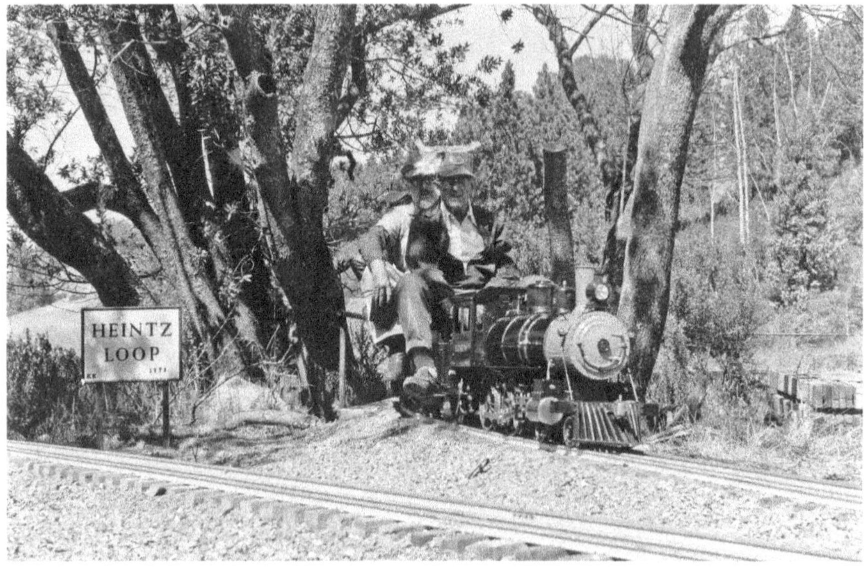

The Golden Gate Live Steamers club constructed the Redwood Valley Railway, an elaborate collection of miniature trains, in the early 1950s. *Courtesy East Bay Regional Park District archives.*

trails and picnic sites. The Redwood Bowmen had hosted the archery range there since 1939.

Inside the main entrance to Redwood Park, the Golden Gate Live Steamers miniature railroad club constructed an elaborate system of trains on different track gauges—the Redwood Valley Railway.

Under Walpole, practical considerations received attention in Redwood Park as well. Roads were paved, particularly from the park entrance and past the old hunting lodge to Trail's End. Wes Adams got a secondhand crash truck from the airport to replace his red Studebaker.[279]

Grass Valley

Above Lower San Leandro Reservoir and Lake Chabot, rolling hills surround Grass Valley. The land had been owned by Anthony Chabot's East Bay Water Company before the EBMUD water district was formed.

Even after the area had been sold by the owners of the original Grass Valley Ranch, it had been leased for cattle grazing for many decades. During the late 1940s and early 1950s, park managers realized the importance of expanding the district southward to serve the people of Castro Valley and Hayward.

Grass Valley Park opened in 1952. No one could be more pleased than the numerous local equestrians and riding clubs. A trail ride was organized, featuring hundreds of local riders, as part of the opening festivities. While the valley and meadow are indeed grass-covered, extensive planting of eucalyptus trees by the early water companies has made an impact on the landscape.

William Penn Mott (1909–1992)

In the pantheon of singular individuals who shaped the open space of the East Bay redwood region, William Penn Mott is a key figure.

Mott, originally from New York, began his career in 1933 as a landscape architect for the National Park Service. He joined the City of Oakland's parks department in 1946. Of his early career achievements, perhaps the best known is the creation of Oakland Children's Fairyland at Lake

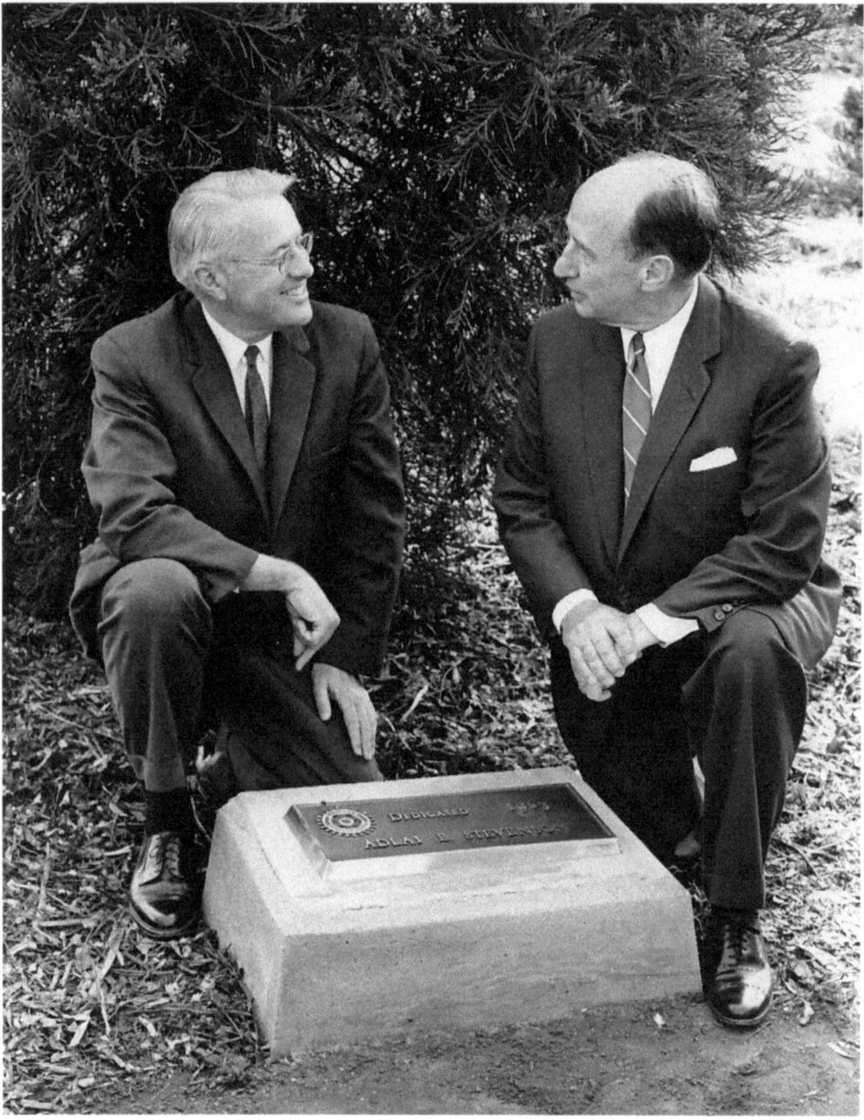

William Penn Mott and Adlai Stephenson, United States ambassador to the United Nations, officiate at the dedication of the Peace Grove on Wildcat Peak, circa 1964. *Courtesy East Bay Regional Park District archives.*

Merritt. The magical array of pastel structures and spaces, evoking beloved storybooks and scaled to those under twelve, is widely credited as the inspiration for Walt Disney to create his original Disneyland.[280]

Mott in the City of Oakland

"Mott was a great guy, always up-tempo in his attitude. He was an all-around fine man," recalled Bob Schultz, who was hired by Mott to build exhibits at Children's Fairyland. "Mott was a more or less a hands-on guy. He assembled a very competent staff of carpenters, plumbers and gardeners. And he hired naturalist Paul Covel....He saw no barriers to getting a project done. He'd say, 'To hell with finances; let's just do it. Send me a bill when it's finished.' It was never a question of 'Well, we'll have to go before the city council and find the money.' He was not one to argue with a committee."

While Mott was a man who appreciated big ideas, he was also exactingly detail oriented. Schultz recalled:

> *Mott made the rounds of all his parks every day, to be sure that everything was shipshape. A gardener was assigned to each park. Each gardener had a tool shed with a coffeepot. Mott would visit each day, taking a special interest in that park and fostering a spirit of competition among the gardeners.... If the gardener would ask for a particular material, Mott would say, "Call Joe to run out to Ogawa's nursery at 73rd Avenue and Foothill." Frank Ogawa was always a great guy for donating material. There was no such thing as a work order. If he saw that the grass was damaged, we'd be cutting out sod to replace it. The lawns in all the parks were always perfect. Mott saw to it.... One time I had driven my pick-up truck down the narrow path at Children's Fairyland, and the tire went off the pavement, just six inches. Mott soon saw the impression of the tire on the fresh grass. The other guys told me, "Schultz, get your truck out of Fairyland! Mott's going to have an investigation!"*

At the same time, Mott was always receptive to suggestions from his staff. One of Mott's grand schemes for Oakland was to dam Sausal Creek above the Leimert Bridge to create a hillside lake, with boating and resort hotels. He also encouraged the Metropolitan Horsemen's Association, then nine hundred members strong, to build an elaborate equestrian center in Joaquin Miller Park.

Mott was, however, a great believer in using private funds to build recreational facilities. The original Children's Fairyland was built with $50,000 raised by the Lake Merritt Breakfast Club. One key fundraiser was Earl Warren.[281]

Mott as Regional Parks General Manager

Jerry Kent told the story of how the East Bay Regional Park District recruited Mott to come to work as general manager in 1960: "Clyde Woolridge and Gordon Sproul went to ask Mott to come to the park district. Mott told them, 'I'm doing a culvert project right now; come back and see me again in two years.' So Wes Anderson, who had been the Redwood Regional Park supervisor during the World War II years, served as district general manager until 1962, when Mott agreed to come."

Mott led the park district from 1962 to 1967. During that time, he hired managers who would make their own mark: naturalist Chris Nelson and Public Information Director Richard Trudeau. At the same time, he was supportive of the employee union, having worked in blue-collar jobs himself.[282]

Mott also felt that the moment was at hand for open space land to be protected before it could be sold to developers. For this, he needed revenue. The elegant solution was to bring Contra Costa County into the park district. This was not a straightforward problem to solve—park bond issues had repeatedly been rejected by the voters east of the tunnel.

Partnering with county counsel (later state senator) John Nejedly and citizen activist Hulet Hornbeck, Mott played a leading role in annexing parkland in western and central Contra Costa County. Briones, Kennedy Grove, Las Trampas and Coyote Hills Regional Parks, as well as the Las Trampas Wilderness, were added to the park district under Mott's leadership.

The Mott era saw the park district land area grow from 7,400 acres to 17,000 acres in five years.

New Priorities Under William Penn Mott

William Penn Mott had different ideas as to how parks should be managed from those of his predecessor. Whereas Walpole had filled Roberts Recreation Area with amenities, Mott saw it as an overcrowded space—and this in the midst of the stunning beauty of the tallest redwoods in the East Bay. In collaboration with Irwin Luckman, Mott ordered some big changes. "Mott said that Roberts Park was too developed," recalled Jerry Kent.

> *First, we sold the merry-go-round to the Hayward Area Recreation District. They sent a crew to disassemble it. The merry-go-round is now in Kennedy Park, on Hesperian Boulevard. Next, they took out the pony rides.*
>
> *Mott said that the twenty-foot picnic tables were too big. So we had a crew take those out. We rebuilt the picnic area, configuring it the way it is today, with smaller areas clustered under the big trees.*

The entrance area to Redwood Park had drawn the attention of Mott as well. In his view, the nearby Big Bear Tavern did not fit in with the character of a park in the woods where families came to enjoy nature. The Big Bear owners, E.D. and Mildred Sellers, sold the tavern, with forty acres of land, to the park district for $83,000 on December 22, 1964. The district obtained an order of condemnation from Alameda County on May 11, 1965, and demolished the building soon thereafter.

Law Enforcement in the Parklands

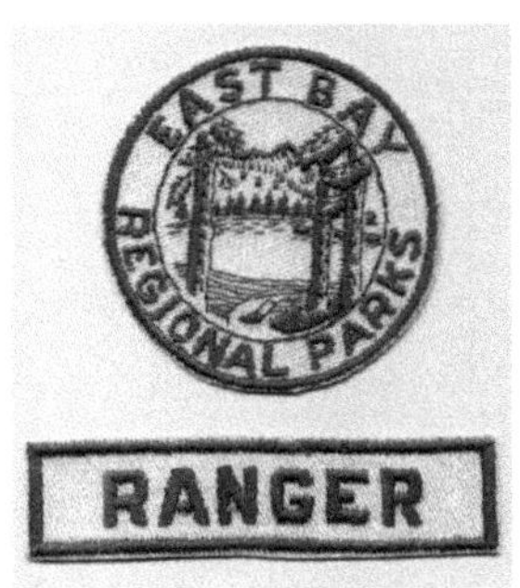

Law enforcement in the park district was provided by volunteer mounted rangers until 1962. Sidney Chown wore this ranger insignia. The rangers also were issued a pistol and handcuffs. *Courtesy East Bay Regional Park District archives.*

Another big change Mott effected within the park district involved law enforcement. At the time the park district was formed, bad actors were brought to justice by volunteer rangers, horsemen who covered many trail miles in the East Bay hills. One of the first was Sidney Chown, along with his associates from the Piedmont Trails Club.

During the 1940s, another ranger was Lou Schaackey, who in 1948 rode with Chown across the stage at Oakland's Woodminster Amphitheater during one of Juanita Miller's theatricals for "Joaquin Miller Day."

During the 1950s, the volunteer rangers worked under the supervision of the Oakland Police Department. Chown continued in this capacity, along with noted horse trainer Jimmy Black, Louis Mustblas, Tom Bacon and Lloyd Graham. The volunteer rangers were issued handcuffs and pistols to carry on their belts.[283]

According to park historian Jerry Kent, when William Penn Mott became the district general manager, he requested that the City of Oakland deploy a regular police force expressly for the parklands. Oakland declined, citing more urgent priorities.

That was how it came about that the park district instituted its own professional police department in 1962. The volunteer ranger force was decommissioned—a temporary situation, as it turned out.[284] "I had to go up to the district headquarters[285] and turn in my gun and badge!" Lloyd Graham Donaldson recalled in 2015. The badge once worn by Sidney Chown remains, a museum piece, in the park district archives, having been donated by his descendants.

The park district police, however, would continue to rely on Volunteer Mounted Patrols through the next half century. In an organization similar

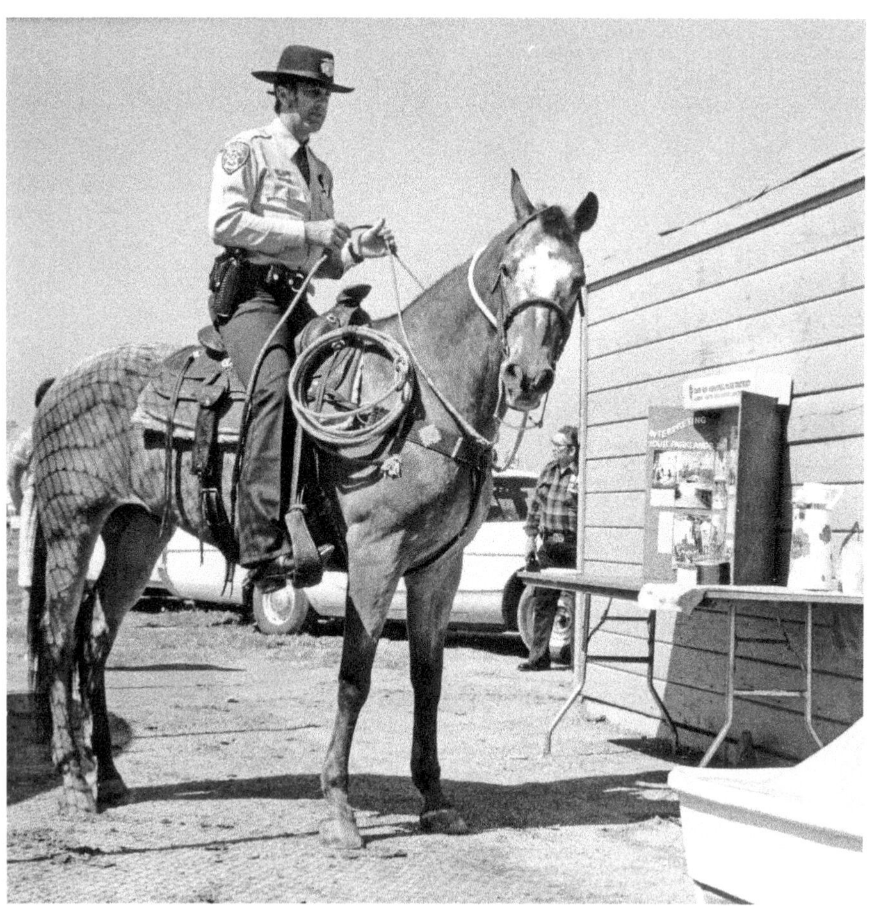

Harry Brizzee was one of the first mounted police officers who began serving the park district after 1962. *Courtesy East Bay Regional Park District archives.*

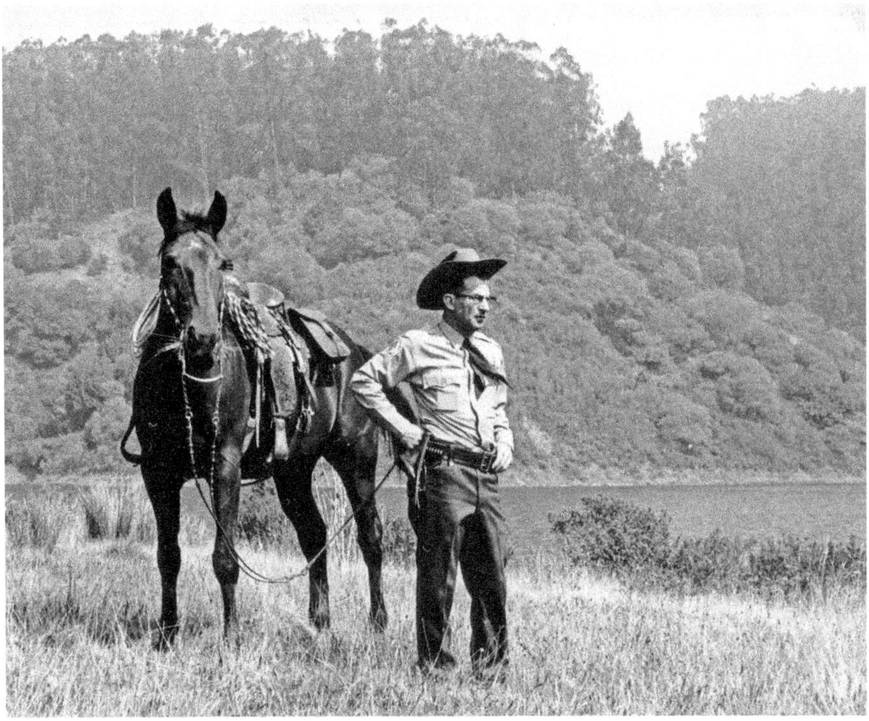

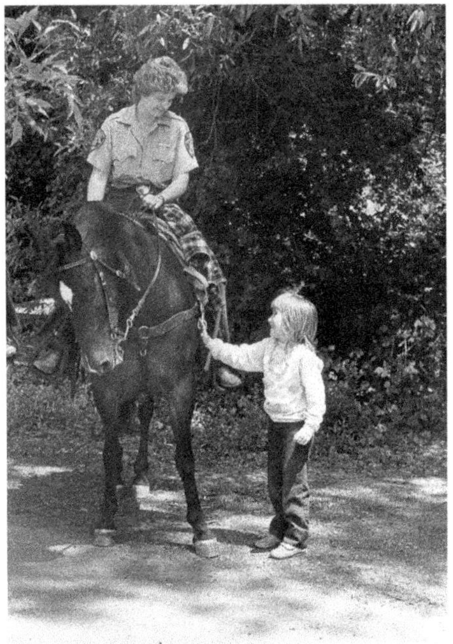

Above: An unidentified mounted police officer stands by Lake Chabot circa 1965. *Courtesy East Bay Regional Park District archives.*

Left: Mounted police officers and volunteers serve as goodwill ambassadors to the public. *Courtesy East Bay Regional Park District archives.*

to neighboring counties' sheriffs' posses, the mounted patrols work under the direction of the mounted officers.

Harry Brizzee, who started in 1975, was one of the first mounted officers employed by the park district. Lloyd Graham recalled Brizzee coming to call at Piedmont Stables during the snowstorm of 1962, happy to have gotten through the snow to partake of hot coffee in the lounge.

In 1979, mounted officer Dana Weaver succeeded Brizzee. Mounted officers who followed included Fred Michael, Malary Hathcox Anderson, Alfonso Anaya, Jim Grabowski, Tracy Desiderio, Ben Guzman and Tom Walsh. All served as liaisons with the Volunteer Mounted Patrol.

After a pattern of user group conflicts emerged between mountain bike riders and equestrians, in 1986 parkland unit manager Renee Crowley began meeting with bicycle user groups. Michael Kelly, who had collaborated with the Tilden-Wildcat Horsemen's Association and East Bay Trails Council, came forward as a leader in what would soon become the park district Volunteer Bicycle Patrol.

JUVENILE INMATES BUILD THE GOLDEN SPIKE TRAIL

A particularly rainy winter in 1962 caused significant flooding, with Redwood Creek overflowing onto the road. A water tank above Piedmont Stables slid down the muddy hill, ending up in the middle of Redwood Road. "In 1962, there was a big storm and flood," Lloyd Graham recalled. "We could not get out to the driveway. There were two inches of mud on the floor of the Piedmont Stables lounge. What a mess! I thought I was out of business. They had to close Joaquin Miller Road."[286]

In the aftermath, park managers decided to build a new trail, parallel to the increasingly busy Redwood Road, on its north side, connecting the Hunt Field by Lorimer's Oakland Riding Academy with the Stream Trail area. Inmate crews from the Alameda County juvenile probation department provided labor. "One crew started above Lorimer's and the other started near the Stream Trail," Jerry Kent recalled. "They met on the trail close to Piedmont Stables."

A monument with a brass plaque stands at the spot where the two crews met in August 1965. The boys' labor created the Golden Spike Trail, so named as a tribute to the first transcontinental railroad builders. In 1869, a team laying track from the west met its counterpart team working from the east at Promontory Point, Utah. There railroad tycoon Leland Stanford hammered in a golden spike.

After Redwood Regional Park's Golden Spike Trail was completed, however, Lloyd Graham noticed that someone had punctured the new water tank above the stables.

Changes to Redwood Regional Park in the 1970s

Despite no access to amenities such as a visitor center or gift shop, the public began to visit Redwood Park in large numbers by the 1970s. The road along the stream, originally paved by WPA crews, was open to automobiles all the way to Trail's End.

"It was a madhouse there on weekends," Jerry Kent recalled. "Cars were driving back and forth on the road. There wasn't adequate parking. In those days, people could rent a horse from the Big Bear Stables and just ride into the park. There was a bad equestrian accident on the road."

Managers decided to close the road to cars between the Orchard picnic area and Trail's End. Unsure of whether visitors could manage to walk farther into the park carrying picnic supplies, managers made a visit to Anaheim. They purchased one of the electric trams used to shuttle visitors around the immense Disneyland parking lots and put it into service on the Stream Trail. After a brief trial period, it was concluded that visitors were willing to hike in after all.

Big Bear Stables was closed and the buildings demolished.

The Redwood Valley Steam Trains, run by the Golden Gate Live Steamers club, was also seen as inappropriate for the wildland park. They were moved to their present location in Tilden Park around 1972. With the steam trains gone, the Wayside staging area was cleared for much-needed parking—including space for horse trailers.

Hulet Hornbeck (1919–2012)

No list of remarkable people in the history of East Bay open space would be complete without Hulet Hornbeck. "[Hornbeck] is one of the great conservationists in California. He's a towering figure. Not only did he know every square inch of Alameda and Contra Costa Counties, he knew

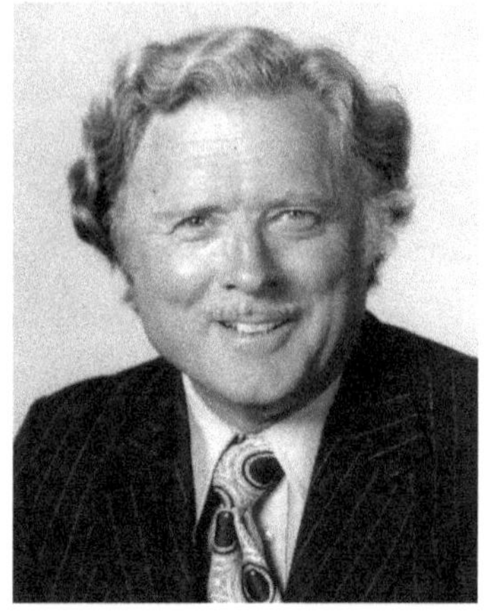

Hulet Hornbeck, the first EBRPD land acquisition manager, overcame cancer while hiking, horseback riding and doubling the parkland acreage. He lived on for thirty-seven more years, passing at the age of ninety-three. *Courtesy East Bay Regional Park District archives.*

who owned it," said park district director Harlan Kessel, on the occasion of Hornbeck's retirement.

In 1963, while Hornbeck was president of the Contra Costa Parks Council, he was approached by William Penn Mott. They agreed to collaborate on an initiative to bring Contra Costa County open space into the park district. In 1964, voters approved the annexation. Shortly thereafter, Hornbeck was diagnosed with melanoma. The doctors estimated that he would have five years to live. "I had a cancer operation," he recalled in his oral history.[287]

> *I was in the hospital, Alta Bates, lying on my stomach, thinking. A melanoma…is like wildfire, supposedly. So I thought, "What do I really want to do?"*
>
> *I was hardly out of my bathrobe when I went over to Bill Mott's house. I had never thought of working for the park district even once before. Not even once….But then, in the hospital I thought, "Well, I had better do it full time if…I have only a limited number of years."*

Hornbeck, who had been an attorney in the insurance business, quit his job and joined the park district as its first full-time land acquisition manager. By 1981, he had played an active role in acquiring all the parkland that Mott had initially lined up. Hornbeck is also credited with acquiring Brooks Island, Black Diamond Mines and the extension of the Lafayette-Moraga Trail for the park district.

Reminiscing about his legacy, in 1981, Hornbeck told the interviewer, "I believe that the last fifteen years or so of this land program has brought us the best park district anywhere in the U.S. in an urban area, during any time in the history of this country."

The doctors had told Hornbeck he had five years to live, but he went on to live thirty-seven years more. After retiring from the park district in 1981, he continued to enjoy his family and his horses until he died in 2012 at the age of ninety-three. The Hulet Hornbeck Trail on the ridge between downtown Martinez and the John Muir Historic Site is named in his honor.

Chapter 9

HORSE PEOPLE

Horses bring out the best in everyone.
—Jan Studley Hansen

Redwood Park Equestrians in the Early Twentieth Century

Trailblazer Sidney Chown and the Piedmont Trails Club

Sidney Vickers Chown (1887–1961) was a celebrated East Bay horseman and grocer. As founder of the Piedmont Trails Club and its namesake, Piedmont Stables, Chown blazed trails and led equestrian sports in the East Bay hills parks. Half a century after his passing, he is remembered as a good friend to all.

Chown, a San Francisco native, came from a pioneer family. His maternal grandfather, George Washington Stilwell, came to California from Pennsylvania in 1849 for the Gold Rush. From an early age, Sid manifested an affinity for two things: work and horses. By 1900, the Chown family was living at 345 Hanover Street,[288] on a hilltop overlooking Lake Merritt. The family kept five cows in the backyard, selling milk to the neighbors. Sid was responsible for driving the cattle to pasture.

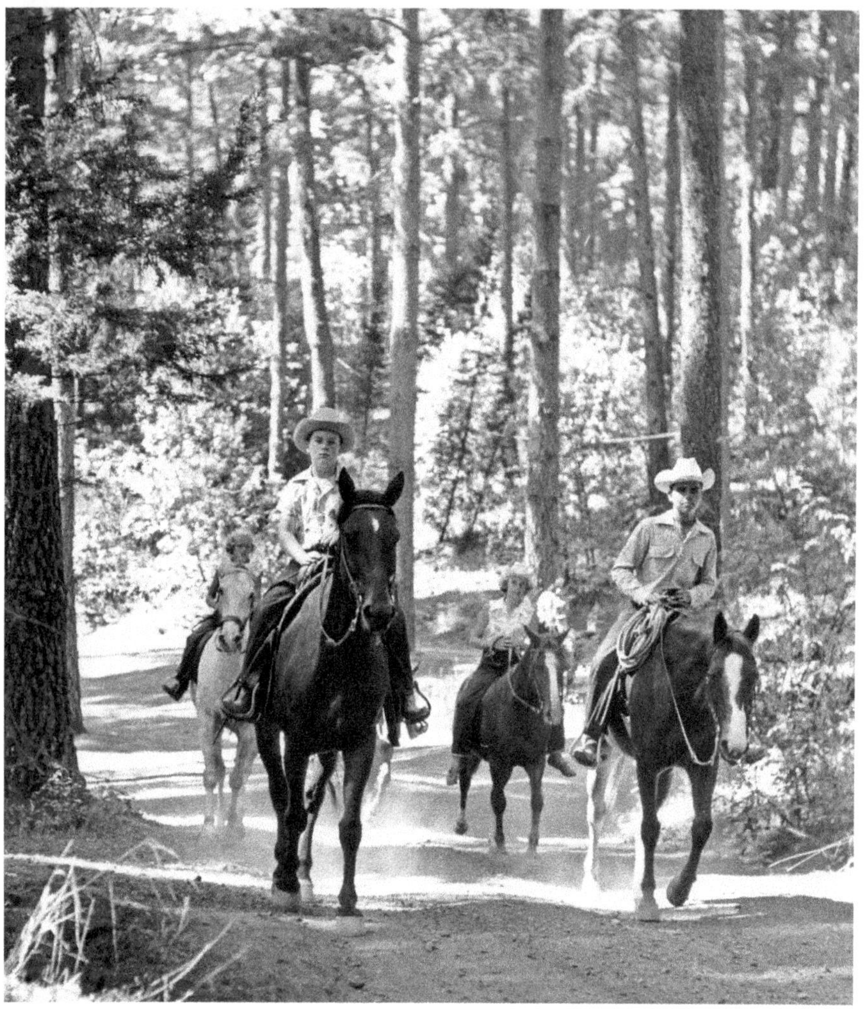

Riders in the redwoods in 1960 include Joanie Peterson, *center*, and Bobby Garcia, *right*. *Courtesy East Bay Regional Park District archives.*

"Borax" Smith Helps Sid with His Cows

Tycoon Francis Marion ("Borax") Smith, whose palatial estate Arbor Villa stood on a nearby hilltop,[289] often passed little Sid and his cows on his daily walks to the Realty Syndicate Building at 1440 Broadway. On at least one occasion, Smith, an erstwhile Wisconsin farm boy himself, helped Sid carry the heavy chains that were used to prevent the cattle from straying.

Sid attended Franklin School through fourth grade and then went to work to help support the family.[290] After school, in addition to watching the cattle, he cared for neighbors' horses. He earned some money doing it, but his primary motivation was to be around horses. Sid saved his money and bought his first pony from a vegetable peddler for fifteen dollars. By the time he was in his early teens, Sid had become a horse trader.

Sid Chown and Joaquin Miller

Around 1902, Sid sold a horse to Joaquin Miller for fifteen dollars at the Hights, the poet's hillside home and literary retreat. Upon closing the deal, Miller gave Sid a drink of wine. The boy felt so elevated that he ran all the way down the Sausal Creek Canyon and back home to Hanover Street. Sid's mother, Jessie, wondered what had gotten into him.

After two weeks, Joaquin Miller told Sid that he was not pleased with the horse; it had taken to eating the bark off the young trees that the poet had planted. Sid took the horse back and refunded the fifteen dollars. Retelling the story in 1961,[291] Sid noted that he prided himself on making honest deals.

At age twenty-seven (about 1914), Sid decided to go into the grocery business.

Sid Chown Buys a Ranch in Piedmont

Around 1918, Sid was riding through the hills into Montclair district of Piedmont when he was hailed by a "real estate fellow": "Hey, there, why don't you buy an acre of land?"

Chown thought that was such a good idea that he bought three acres, surrounded by ancient oak trees, located at 2220 Andrews Street, between Mountain and Snake Boulevards.[292] Around the same time, Chown married his first wife, Florence Cobbledick, who came from a distinguished Oakland family. Their home during the first year on Andrews Street was a tent.[293]

Soon Sid was operating two grocery stores. According to *Polk's Oakland City Directory, 1922*, his markets were at 2901 College Avenue in the Rockridge district and 1555 Solano Avenue in Berkeley. Gradually, the Chowns built up their ranch. First, there was a one-room house and a small stable. Gradually, more rooms were added to the house, and by 1923, there were twelve stalls for horses.

Chown and his friends organized the Piedmont Trails Club. Extra stalls at 2220 Andrews were rented to club members. They went on weekly hikes and rides and, in so doing, discovered there was a great need for more trails in the vast open space of the Oakland hills. They set to work constructing trails by hand.

As the Chown ranch evolved, the residence was built over what had been the foundation of the stable. By the mid-1930s, only a few stalls remained.[294] From about 1922 through 1931, Sid worked for the Caterpillar Tractor Company. His associates there were said to be "very horse-minded."

In 1925, Sid put on the first Oakland Horse Show at the Oakland Civic Auditorium. To publicize the event, he went around to schools, telling all the children.

In July 1931, the Caterpillar Tractor Company closed its Oakland operation and moved back to Peoria. Sid lost his job. While Sid was "between jobs," he single-handedly, using a machete, blazed the Piedmont Trail in Redwood Regional Park, connecting the (Harold) French Trail to the Stream Trail in Redwood Canyon.

By 1934, Sid had decided to go back into the grocery business.[295] He opened a market on Leimert Street, and he also a much smaller one on Mountain Boulevard, just down the street from the family ranch. There he set Florence up in business.[296] Their marriage had ended in divorce.[297]

Granddaughter Heather Galanis recalled that the tiny market also had an old-fashioned double-gravity gasoline pump with an elevated glass cylinder. "As a young child, I used to watch with fascination as the pink gasoline surged through the glass as people pumped it."

Sid remarried. According to Lucile Chown,[298] she and Sid opened a small grocery store on Leimert Boulevard in 1934. It prospered, and they moved across the street to a larger building at 1446 Leimert[299] in 1937. "My grandfather had a rough-and-ready style," Heather observed. "When he said he 'had to see a man about a horse,' it usually meant a trip to Livermore."

"Time and time again, [Sid] would go out of his way and scout around to find a mount for those who needed one or expressed a desire to locate a horse better suited to his or her needs. He did not expect or would he ever have accepted any remuneration for this service," wrote George Daniels, Sid's friend from the Metropolitan Horsemen's Association.[300]

Sid became one of the local horsemen who served the East Bay Regional Park District as volunteer rangers. Wearing badges issued by the park authorities, they would pack picks and shovels on their horses to maintain trails in the backcountry.

Sid joined many organized horse clubs. In addition to the Oakland-based Metropolitan Horsemen's Association (MHA), he rode with the Frontier Boys, the Sonoma County Trail Blazers, the Reined Cow Horse Association, the Mount Diablo Trail Riders, the Merced-Mariposa Cow Horse Association and the California State Horsemen's Association.

By this time, Sid was very well known around town. He worked with the William Penn Mott and the city parks department to construct the Sequoia Horse Arena in Joaquin Miller Park, which has been in continuous operation by the MHA since it opened in May 1948.

One lasting impact of Sid's public advocacy is the horse trail that traverses the hillside median for several miles between the northbound and southbound lanes of Skyline Boulevard south of Redwood Road.[301]

In 1949, Sid posed for a photograph, mounted on Guy Diablo, in front of the equestrian statue of Joaquin Miller, a local landmark. Chown told the *MHA Trail Blazer* that the statue marked the very spot where he, as a boy of

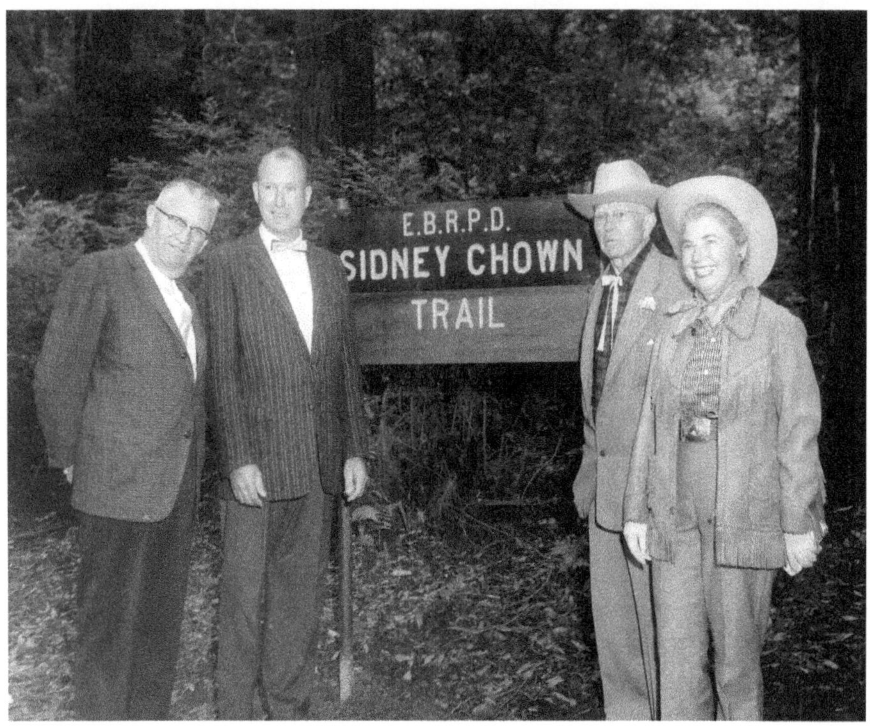

Park district officials John MacDonald and Wes Adams congratulate Sidney and Lucille Chown in 1961 when the Piedmont Trail that he blazed was renamed in his honor. *Courtesy East Bay Regional Park District archives.*

thirteen fifty years before, had sold a horse to Joaquin Miller. Chown noted that the statue was at the site of Joaquin Miller's old corral; at that time, there were no trees on the grassy hills. It was Miller, along with "Borax" Smith, who planted the trees that now fill the parkland, he said.

Sid and fellow ranger Lou Schaackey made the news when they rode their horses across the stage at the Woodminster Amphitheater on Sunday, September 25, 1949. The occasion was one of the theatrical events for "Joaquin Miller Day" staged by Juanita Miller, daughter of the late poet. Juanita played "Kerats, a Waif of the Mining Camp." The mission of Chown and Schaackey was to rescue "actress and pioneer equestrian Inez Thompson from 'hostile Injuns.'"

"Sidney Chown was such a nice man," recalled Inez Thompson Fort, longtime horsewoman and resident of the Joaquin Miller Park Heights neighborhood.

In 1954, Sidney was awarded the East Bay Regional Park District President's Cup, presented by Director Robert Sibley. The year 1954 also marked Sid's retirement from the grocery business.

By 1959, Sid was ill with cancer. It did not stop him from again joining the Yosemite Trail Ride, this time winning second place. Weakened following surgery, he could not saddle his horse or mount without assistance. Still, he completed the strenuous three-day 1959 Frontier Boys trail ride in heavy rains. "I rode much of the way with Sid, and I can say he put many of the younger fellows to shame," his friend George Daniels wrote.[302] "Once in the saddle, he was as good as any when the going got tough."

Sidney Chown died in Oakland on December 12, 1961. He was seventy-three.

Horse Barns of the 1930s

By the time the EBRPD began buying up private land in Redwood Canyon in the 1930s, hundreds of local residents kept horses there. According to the *Trail Map of Redwood Regional Park and Its Environs*, compiled for first park district general manager Elbert Vail, at least ten stables lined Redwood Road between Skyline Boulevard and the park entrance.

Piedmont Stables

Piedmont Stables was moved from the Chown Ranch in Piedmont to Redwood Canyon in the 1930s. The Piedmont Trails Club needed more stall space. The site had been the location of the dairy run by the Ramos family, whose daughter Lydia married John Reis.

Over the next twenty years, the sixty-five-horse stable became popular with affluent equestrians from Piedmont. A crew of grooms was employed so that the well-heeled ladies could enjoy the trails without soiling their riding habits.[303]

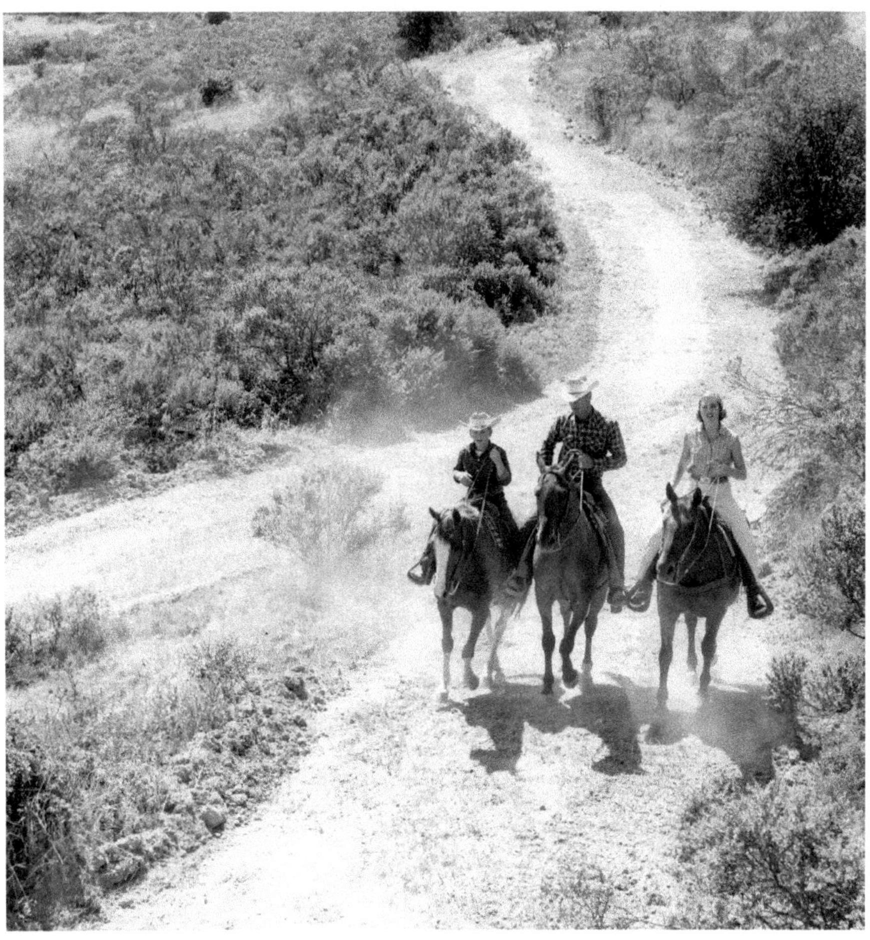

Lloyd Graham Donaldson, longtime operator of Piedmont Stables, rides with his wife, Rosamund, and son, Jeff. *Courtesy East Bay Regional Park District archives.*

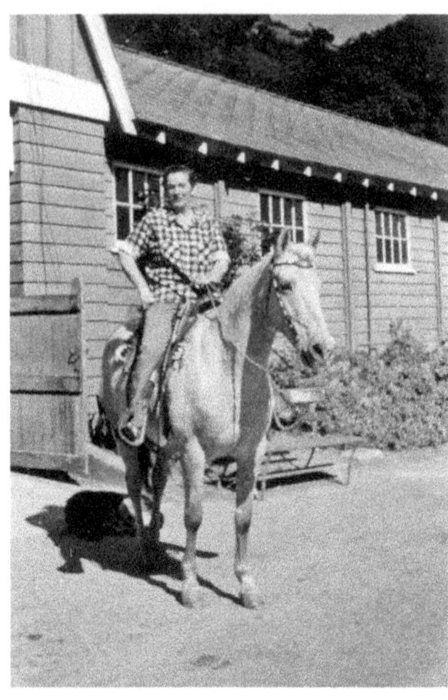

Unidentified rider in front of Piedmont Stables, circa 1955. *Courtesy East Bay Regional Park District archives.*

The origins of the name of the Tate Trail, which ascends a ridge east of Piedmont Stables, are not known. Older maps show it as the Charles Tate Trail. Since there are no documented landowners of that name in the vicinity, it is likely that Charles Tate was a member of the Piedmont Trails Club.

Other Barns in Redwood Canyon in the 1930s

Near the southeast corner of Skyline and Redwood, beloved British horsewoman "Miss" Beatrice Graham was then operating her third riding academy, at the site of today's Skyline Ranch. Across Redwood Road on the northeast corner stood the Redwood Riding Academy, which Miss Graham would purchase during the 1940s. Even after she became Mrs. Donaldson, she was known throughout the horse community as "Miss Graham." Her son, Lloyd Graham Donaldson, was similarly known as "Lloyd Graham." The Graham Trail in Redwood Regional Park is named in honor of their family.

Proceeding downhill on Redwood Road, the Oakland Riding Academy, owned and operated by Bob and Lorraine Lorimer, served a clientele of affluent hunt-seat riders, in a friendly competition with Piedmont Stables. A smaller barn run by the Rennaeker family was on the east side.

Farther down, at the side road leading to the Big Spring, the Montiero Ranch and the Roundup tavern, the B-S ("B-bar-S") ranch straddled Redwood Creek. Two smaller barns, Hopalong and Roundup, were also above Piedmont Stables. On the south side of Redwood Road, behind the Big Bear Tavern, the small Canyon Stables stood at the base of the trail leading up to the spring used by the Alhambra Water Company. The rustic Big Bear Stables was across the road from the tavern of the same name, near the Redwood Gate and park office.

In those days, Redwood Road was not the thoroughfare it would become after the 1970s. Visitors piled into hay wagons for evening rides, with a red lantern swinging at the tail of the wagon. Automobile traffic was light. The road twisted around curves. In stretches where the road widens today, the shoulder was the original road curve before the route was straightened.

With several taverns in the vicinity, however, trees along the roadside had been painted white to reflect headlights. Photographs from the 1930s show the painted trees; some near Piedmont Stables today are still marked with the reflective paint.

On the west-facing slopes of old Redwood Road in the 1930s, only the Wildwood Ranch stood, on the north side. Heading toward Leona Canyon on the Windy Gap Trail, the rider would pass the Diamond-T Ranch, which would later be purchased by Oakland auto dealer Raymond Miller and his wife, Inez.

Other points of interest are shown on the 1930s-vintage Elbert Vail map. Lodging in the canyon for hunters and other visitors could be found at the Redwood Hotel, a short distance east of the Redwood firehouse. The Redwood Hunting Lodge was also still in operation, until the coming of the park district in 1939.

Horse Barns of the 1950s and 1960s

By 1950, Miss Beatrice Graham was operating her Redwood Riding Academy at the northeast corner of today's Skyline Boulevard and Redwood Road. She lived in a cabin by a large spring next to her riding arena. She died there in the early 1950s.

Lloyd Graham had returned from service in the U.S. Marine Corps. He continued to operate Miss Graham's Redwood Riding Academy until 1954. At that time, the property was sold for development. Lloyd accepted an offer to manage Piedmont Stables, where he remained for the next half century.

Bob and Lorraine Lorimer in the late 1940s purchased the Oakland Riding Academy, serving English-style hunter/jumper riders. In the early 1960s, the Lorimers collaborated with the park district to develop the Hunt Field in the hills to the north of their stable. It included an Olympic-size arena, plus a cross-country course, with jumps, for English-style fox hunting. Its grand opening ceremonies on March 1, 1964, included civic officials, equestrian celebrities, the Los Altos Hunt Club and even a bagpiper.

Across from Lorimer's, Stanley Cosca built Skyline Ranch in 1948. It was intended to be a very modern facility catering to western-style riders.

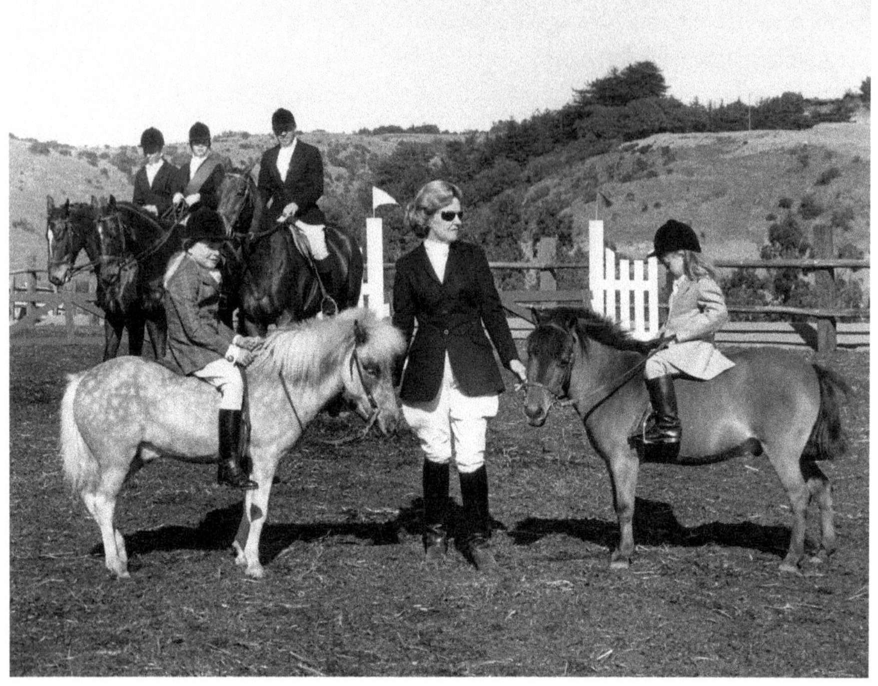

Lorraine Lorimer and two of her daughters celebrate the grand opening of the Hunt Field on March 1, 1964, in upper Redwood Regional Park. *Courtesy East Bay Regional Park District archives.*

The barn featured a concrete floor and individual watering basins in each stall. Trick riders, including young Loretta Cosca, could be seen practicing every day. Steve Cosca, son of Stanley, went on to fame as a rodeo rider and Hollywood stuntman. Celebrated horse trainer Jimmy Black and his wife, Tiny, lived on site, where they raised their three sons. Scotty Black, brother of Jimmy, had his own ranch a short distance north on Skyline Boulevard.[304]

Downslope on Redwood Road, changes were afoot. In the early 1970s, construction was underway for the regional Bay Area Rapid Transit system. A new campus for Merritt College was built on Portuguese Flat above Leona Heights. Redwood Road was widened in the flatlands and straightened at the top of the hill. The Pinto Ranch was demolished, to be replaced by the Pinto Playground and the Carl Munck School. The Wildwood and Wagon Wheel Ranches were demolished for the construction of housing subdivisions in the old Stauffer mining area.

Grass Valley/Anthony Chabot Regional Park

Men's Riding Clubs

Grass Valley, dating from the time when it belonged to the East Bay Water Company, was the playground for a number of midcentury men's riding clubs.

The Aahmes Shrine mounted patrol was best known for its Wild West shows and parades. With Fred Bemis as captain, they kept as many as thirty black-and-white pinto horses at their Rancho San Antonio at 13560 Skyline Boulevard, on a hilltop west of Grass Valley. The name honored the idea that the site had been used in older times by *vaqueros* from the Antonio Peralta Rancho.

Prominent local members included Earl Warren and Kenneth E. Bemis, who had a large ranch at the top of Grass Valley Road, near today's Clyde Woolridge staging area into Anthony Chabot Regional Park.

In the early days of aviation, pilots carrying the U.S. mail depended on a hilltop beacon, located by the Aahmes Shrine stables, for navigation. The Shriners proudly ensured that the signal light was kept in operating order.

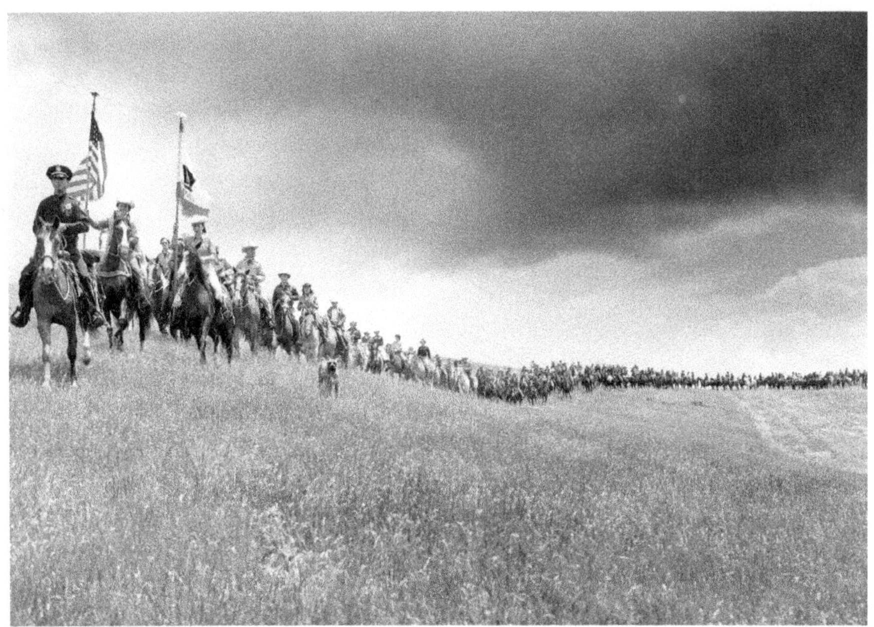

Hundreds of equestrians came for a trail ride as part of the grand opening ceremonies for Grass Valley Park in 1953. *Courtesy East Bay Regional Park District archives.*

The Frontier Boys were another riding club. Participants included Sidney Chown, Claud Brandon, Lloyd Graham and John O'Hanneson, a dentist whose ranch was on the western hillside of Leona Canyon. The Frontier Boys were fond of traveling to the Sierras, or even to other western states, for well-attended trail rides.

Anthony Chabot Equestrian Center

In the years after World War II, the human and equine population soared in the East Bay hills. On the bay-facing slopes of Redwood Road, a new stable called the Wagon Wheel, located at 4951 Redwood Road, had been built on the west side of the older Wildwood Stables. On the south side, where the road took a corkscrew path above the declivity containing Raymond and Inez Miller's ranch,[305] Harold Cummins and Stanley Cosca built the Pinto Ranch.[306] The Redwood Road barns rented horses for trail rides and had western-style events. The Pinto Ranch even had a hot dog stand, hosted by Gloria Isola, a member of the Lari-ettes drill team.[307]

John Sutter, who served as an elected EBRPD director from 1996 to 2016, recalled the Pinto Ranch as it was in the postwar era. "We would ride in the hay wagon from the Pinto Ranch, which was located on Redwood Road, between Mountain Boulevard and Skyline Boulevard," he said. "This was 1945 or 1946. There was no development up there; it was very rural, with nothing but horse trails….I remember that the barn at the Pinto Ranch had bats. There were living bats in the barn, and they also had dead ones nailed up and displayed outside the building. The Pinto Ranch was there into the 1960s."[308]

Yvonne and George White were the barn bosses of the Wildwood Stables. In the early 1950s, according to Yvonne White, they were given notice by their landlords that the property was to be developed to build the Crestmont subdivision.[309]

The Whites called on officials at the park district. Grass Valley had recently been opened to the public, and the trails were very popular with equestrians—shouldn't there be a modern stable and arenas in the new park? Park officials agreed. A deal was struck: the district would provide the land and the materials for the barn and arena. The Whites would provide the labor to construct the facilities.

Architect Irwin Luckman designed a round barn. Each stall had an adjoining outdoor turnout space so that horses could choose to be indoors or outdoors. From an aerial view, the barn and its outbuildings had the shape of a tortoise.

Two large arenas were constructed, with labor provided by teenage boys from Alameda County Probation Department "honor camps." At the end of the workday, the boys were rewarded with an opportunity to ride horses there.

After Yvonne White retired in the 1990s to move to South Dakota to be with her son, George White Jr., a professor of geography, Sara Crary became barn manager. She handed over the reins to Mona Ellingson in about 2010.

Horses in Leona Heights

The historic Leona Heights district of Oakland was once known as Laundry Farm Canyon. It was a rail destination for picnics. The Stauffer Chemical Company had extensive diggings in the Oakland foothills south of Redwood Road. Sulfur, rhyolite and even some hematite were extracted.

Among the Oakland park amenities of the early twentieth century was the Leona Lodge. With its massive fireplace and complete kitchen, the lodge was a favorite stop for equestrians at the end of a ride.

The Old York Trail

What was the route of the old York Trail? Only a few segments remain. According to the late Erling Hansen, the trail was named for a police sergeant York, who was known to be a difficult character.

Starting from the Leona Lodge, the trail once wound up through the mining district, following Horseshoe Creek downslope from the treacherous Devil's Punchbowl, a deep mining shaft. Local lore holds that Native people once extracted blood-red hematite, a ferrous metal used in pigment, there.

Traces of the York Trail can still be followed through the canyon below Merritt College, where "Old Survivor"/"Grandpa" grows. Before the college was constructed, beginning in 1970, the York Trail crossed Portuguese Flat and continued along the gradient of the hill, above where the college running track stands today.

Longtime Oakland equestrian Judi Bank recalled that the York Trail used to follow along the upslope, east side of Sugarloaf Hill, near the college environmental science department buildings. The remainder of the York Trail proceeded south above the eastern side of Leona Canyon, according to Lloyd Graham. It continued along the Rifle Range Branch of Arroyo Viejo Creek to the gun range that was once used by the 143rd Field Artillery of the California National Guard, as well as by the occasional drunken visitor to the nearby Hilltop and Red Barn dude ranches. The southern end of the York Trail was obliterated by the early 1940s, when the Oak Knoll Naval Hospital was constructed there, on Rifle Range Creek.

Early Residents of Leona Canyon

The history of Leona Canyon is tied in with the history of local equestrians. Spanish pioneers originally named it for the mountain lions that made it their territory. Cattle were grazed on the canyon slopes and watered in Rifle Range Creek.

Since 1900, the area had been popular with hunters. Kenneth E. Bemis, a member of the Aahmes Shrine Rangers, kept a hunting lodge and a number of cabins. Bemis, who was in the restaurant supply business, had a partnership with a Mr. White, who ran the White Log Coffee Shop chain of twenty-eight East Bay diners.[310]

A stone fireplace is all that remains of the Bemis hunting lodge, on a hill above the Leona Canyon Trail, northeast of its intersection with the Pyrite Trail. Nearby, on the creek bank, Bemis had a wishing well.[311] "The well had a little roof," recalled Lloyd Graham.

Graham recalled riding his horse through Portuguese Flat and down "Windy Gap," the narrow slot at the top of the Leona Canyon Trail, to visit friends there.

Another Leona Canyon resident was Dr. John O'Hanneson, an Alameda dentist who was a member of the Frontier Boys riding club. His horse ranch was located on a flat area along today's Pyrite Trail. The view from the house site extends southwest over the Coyote Hills and San Francisco Bay. "O'Hanneson was a great guy, a big rugged man," Lloyd Graham recalled. "I remember on one of our Frontier Boys rides, up near Weed, California—he always took his fishing rod along. It was cold! But there he was, out in the lake, fishing." A stub of the old access road to the ranch is labeled O'Hanneson Road on park maps.[312]

Retired park district director John Sutter, a former judge and Oakland city councilman, hiked in Leona Canyon in the 1960s. "I primarily knew

Leona Canyon from Sunday hikes led by Harold Ireland," Sutter recalled. "We hiked there, starting from the Leona Lodge. We would hike up the [Horseshoe] creek trail into Leona Canyon....I was amazed to see the rural houses there. People kept animals—one person was even raising pigeons."

Urbanization Approaches Leona Canyon

After the Oak Knoll Naval Hospital was built, the city gradually crept up to the south side of Leona Canyon. With the construction of Interstate 580, the old Leona Stables, which had stood near the Keller Avenue freeway overpass, was demolished. Leonora Studley, the Leona Stables operator, went to live with Kenneth and Claire Bemis at their ranch at the end of Skyline Boulevard and Grass Valley Road.[313]

Alameda County Flood Control had constructed berms and a pond to impound runoff from the creek at the canyon entrance. The extensive Canyon Oaks housing subdivision was built near the intersection of Keller Avenue and Campus Drive. Soon a corporation proposed to build a subdivision within Leona Canyon.

Leona Canyon Becomes Parkland

Some local residents strongly objected to the idea of development in Leona Canyon. In particular, native plant advocates Marian and Roger Reeve decided to take action to preserve the open space. Marian Reeve, a community college professor of botany, had been a founding member of the California Native Plant Society and was named a CNPS Fellow in 1992. She was also a politically appointed member of the EBRPD Park Advisory Committee (PAC). "Marian Reeve worked behind the scenes with the city of Oakland to preserve Leona Canyon as open space," Jerry Kent noted.

Park director Harlan Kessel arranged with Oakland for the EBRPD to acquire Leona Canyon. The District Master Plan, adopted on May 17, 1988, included strategic guidelines fir future management of the Leona Canyon Open Space.

Meanwhile, within Leona Canyon, a few residents remained. By the 1980s, Kenneth E. Bemis had died. His widow, Claire, and her elder sister were living in a house on the eastern hillside of Leona Canyon. Park staff, including rangers Dee Rosario and Annie Kenny, helped the elderly women with grocery shopping so that they could remain in their home. After Claire became ill and was taken to a nearby hospital, Kenny sat by her bedside.[314]

The O'Hanneson Ranch was inherited by Kimberley McKell. She sold it to the park district in 1998. This was the final property in Leona Canyon to be acquired by the park district, completing the protected open space. The $410,000 purchase price was covered by funds from bond measure AA, passed that year.[315]

"Park district policy for what to do with existing structures within newly acquired parkland is based on several criteria," Jerry Kent explained. Structures are evaluated as to their historic value, their functionality for reuse for park purposes, their accessibility and whether they would be targets for vandalism.

Neither the O'Hanneson horse ranch nor the Bemis house were deemed worth saving. After making a photographic record of the home, park staff demolished all the buildings, leaving Leona Canyon as nature had formed it, except for three trails and the old stone fireplace from the Kenneth Bemis hunting lodge.

Along the Leona Canyon Trail that connects from Keller Avenue to the Merritt College Parking lot, examples of native plants are tagged with numbered signs. A brochure created by Beverly Ortiz, PhD, identifies the native plants and explains their traditional uses by indigenous people.

Leona Canyon has become a popular route for hiking tours for visitors who seek to approach "Old Survivor" from the remnant York Trail.

Chapter 10

NATURALISTS

I am the spirit of a land alive
Upon whose slopes new hopes arise
Green in the fog kissed winter and fall
Giving breast to redwoods tall.
—Louis ("Digger") O'Dell, Joaquin Miller park ranger, 1966–79

Prior to the 1970s, the central motivation for acquisition of parkland by cities and by the park district was recreation—providing restful places for working people to take their families. With increasing public awareness of environmental protection issues, however, a shift in thinking emerged.

By the 1990s, park managers felt a need to balance recreation with resource stewardship. This dichotomy would become an ongoing theme in park policy considerations in the years to come.

Paul Covel: Teaching Oakland's Children about Nature

In 1948, when Wiliam Penn Mott led the Oakland Parks Department, naturalist Paul Covel became his best-known hire.

The local legend of Paul Frederick Covel (1908–1990) endures decades after his long career in Oakland parks. Although he made his headquarters at Lake Merritt, where in 1953 he designed the Rotary Natural Science Center, Covel knew the natural environment in every park in Oakland.

Following the inaugural Earth Day in April 1970, the City of Oakland recognized Covel for his work. "Paul Covel, in my mind, has contributed more to the conservation consciousness of the people of the Eastbay than any other single individual," commented William Penn Mott in 1970, having then become director of the California State Department of Parks and Recreation. "Covel was the first naturalist to be hired by a city park and recreation department in the United States."

After mining operations closed down at the Devil's Punchbowl, Covel led tours of visitors to explore the area.[316]

Fellow naturalist Rex Burress painted a portrait of Covel teaching a nature lesson to three children. The painting hangs above the fireplace at the ranger station in Joaquin Miller Park. "I made the painting from a photograph of Paul in the fields above Leona Canyon where Merritt College now is located. I added the kids to help with the interpretive story of Paul," Burress noted in 2016.[317]

"Rex made those children up for the painting, but so many times I saw that scene enacted by real live adults and kids," commented Covel colleague Rich Wirkkala.

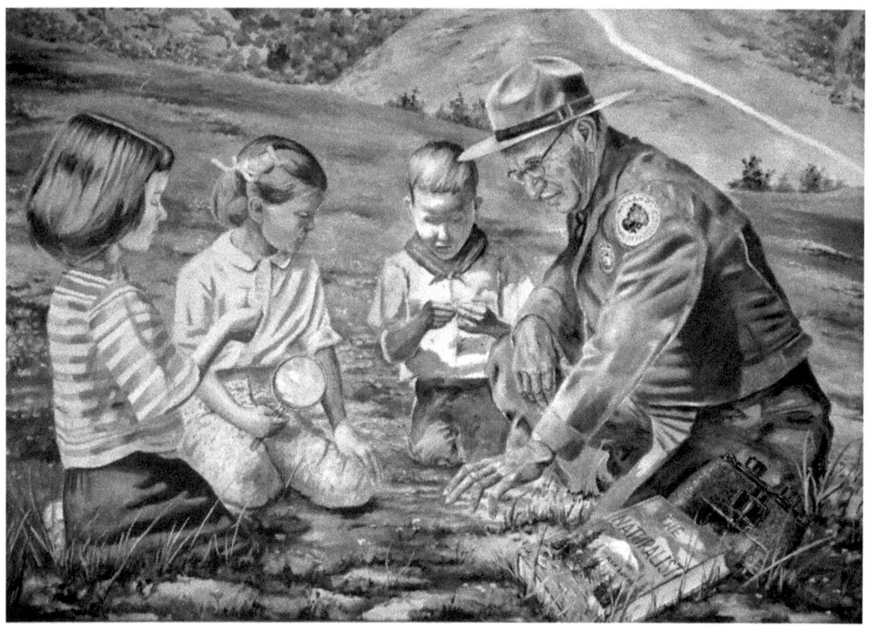

Oakland parks naturalist Paul Covel was beloved by generations of children for his kind manner and his nature lessons. *Courtesy Rex Burress.*

Covel, who lived in the Redwood Heights neighborhood with his wife and four children, was the person credited with the discovery of "Old Survivor."[318] Two of his protégés went on to careers in the Oakland parks: naturalist Stephanie Benevides and Joaquin Miller Park public works supervisor Martin Mataresse. Son Jim Covel became the senior manager of guest experience at the Monterey Bay Aquarium.

Paul Covel was the author of numerous books and articles:

- *People Are for the Birds: Adventures with the First Municipal Park Naturalist at Lake Merritt, America's Oldest Waterfowl Refuge* (Oakland, CA: Western Interpretive Press, 1978)
- *Beacons Along a Naturalist's Trail: California Naturalists and Innovators* (Oakland, CA: Western Interpretive Press, 1988)
- "The Waterfowl of Lake Merritt, Oakland, California" (date and publisher unknown)
- "Trees, Shrubs and Perennials of Lakeside Park" (date and publisher unknown)
- "A Walk in the Coast Redwoods of the Oakland Skyline: The *Palos Colorados* of Rancho San Antonio" (unpublished, circa 1986; available in the EBRPD Archives)

The rich elements of nature that Covel could see are reflected in a stream-of-consciousness account from his article on the *Palos Colorados*:

> *California huckleberry, Heartleaf Manzanita, Redwood Rose, blackberry and thimbleberry, occasional Salal and wax-myrtle are understory shrubs of our Skyline redwoods, with toyon, madrone and Coast Live Oaks around forest edges and clearings. Birds and mammals are fewer than in surrounding plant associations. Noisy Stellar Jays and bark feeders like the Brown Creeper and the Red-breasted Nuthatch forage in redwoods; deer browse through, raccoons, Grey Fox, weasels and skunks find food in the humus and rotting logs on the ground. Salamanders, garter snakes, alligator and blue-bellied lizards are fairly common. An occasional dampwood termite colony may be found under the bark of a fallen log.*

THE LUMBERING HISTORY FESTIVALS

Park district employees who spend any time amid the redwoods inevitably develop a fascination for the lumbering history of the area. In 1984, forester John Nicoles envisioned a program for schoolchildren and the public to celebrate the logging era. "It was the fiftieth anniversary of the park district, and special events were being held in different parks to celebrate. I was asked to create a history program," Nicoles recalled in 2017. "In Redwood Regional Park, the history of the area is one of logging old-growth trees. We decided to have a public program on Columbus Day weekend, to educate the public about how logging technology evolved."

Recalling the hardworking, hard-drinking, oath-bellowing ox drivers of the 1850s, Nicoles decided to call the event the Bullwhackers' Jubilee. The event was held in 1984, 1985 and 1986. "Some people objected to calling it the Bullwhackers' Jubilee because that did not seem like a suitable thing to be teaching children," Nicoles commented. "But in fact, the loggers did whack the oxen, and they were called bullwhackers. There was an historic tool, used to roll logs onto and off of rail cars, known as a jackscrew. I suggested that if there was an objection to the 'Bullwhackers' name, we could call it the Jackscrewers' Jubilee. Objections were dropped."

Reaching out to his professional network, Nicoles arranged to bring actual lumber industry equipment:

> *For these programs, I brought together some real logging equipment such as has been used in redwood logging on the North Coast from the 1860s to the present. During the logging era in Redwood Regional Park—mainly in 1842–55—they used pit saws, where one man stands in a pit below the timber, one man stands above and they use a cross-cut saw. They sometimes used water-powered circular saws.*
>
> *These primitive methods were not conducive to illustrate a comprehensive program, so we did a time line approach. We set up a trail through the woods, where the visitors would pass photographic displays, supplemented with examples of real logging equipment.*
>
> *One year I brought in a steam donkey—that is a steam-powered winching engine used in logging—and operated it. I borrowed a locomotive—it went off the tracks, and the $250,000 Peterson tractor was used to pull it back on. Another year I borrowed a portable sawmill and cut some salvaged logs into lumber. There were also two bull teams to pull logs. We had many conversations with the public that conveyed educational content.*

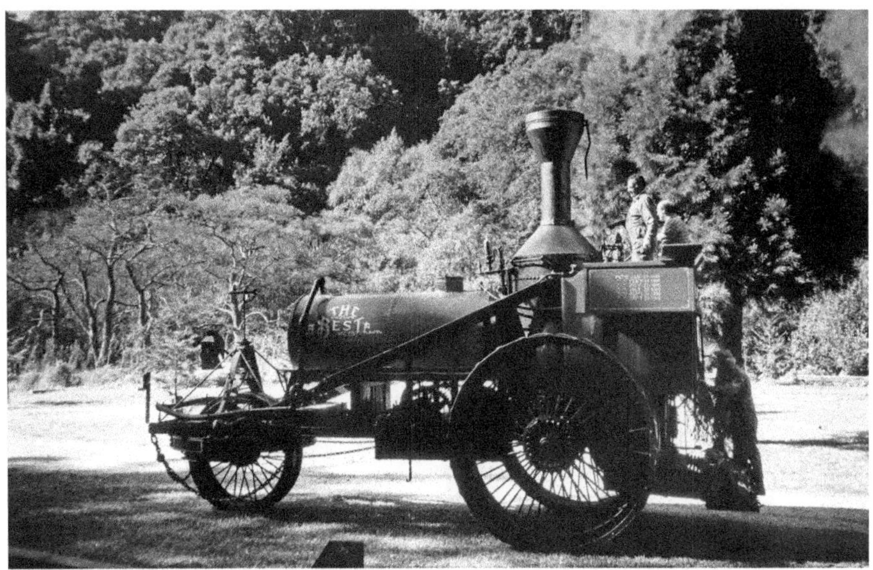

The Bullwhackers' Jubilee festivals in Redwood Regional Park celebrated the logging history of the area from 1984 to 1986. Forester John Nicoles brought in antique lumbering equipment. *Courtesy East Bay Regional Park District archives.*

> *Nine busloads of children were brought in from underserved schools. We gave them a guided tour, led by volunteer forestry professionals, in a loop through the exhibits. We pointed out things and talked about what equipment was used for. We started out with logging and then taught the kids about some of the agricultural activity in the Stream Trail area after the area had been cut over. For example, there are still apple and pear trees growing in the Orchard area that were planted by* [John Reis and his family]. *Dreyer's donated ice cream, so each child got a cup. One volunteer demonstrated how pencils are made and gave a pencil to each student.*

But in the third year of the Bullwhackers' Jubilee, park district naturalists took issue with the program content. The interpretive services department decreed that the Bullwhackers' Jubilee had to stop. These critics made two arguments: first, an exhibit of logging machinery did not honor the ethic of stewardship; second, the twentieth-century equipment on display was actually not a part of the Redwood Park history. "The naturalists said the event was too much about logging," Jerry Kent commented. In 1986, the name of the event was changed to the Redwood Forest Festival: Historical Logging and Forestry Exposition.

Stewardship of the Redwood Creek Rainbow Trout Population

Back in the 1855, when William P. Gibbons discovered the rainbow trout in Redwood Creek, the waters and these native steelhead ran unimpeded to San Francisco Bay.

From its headwaters springing from serpentine rock, this tributary of San Leandro Creek pours down through three reservoirs, where it has been confined, since Anthony Chabot built his dam in 1875. The rainbow trout of Redwood Regional Park have survived the logging era, subsequent sport and sustenance fishing and the coming of the park district.

In the 1970s, the district created an unintended threat to the small residual population of genetically unique trout. This took the form of a dam blocking the creek where it had been bridged to provide access to a fire road.

Within ten years, wildlife biologists discovered that the trout population was declining. The fish ladder that had been built over the dam was insufficient for the migrating fish to climb a five-foot barrier to reach their spawning grounds.

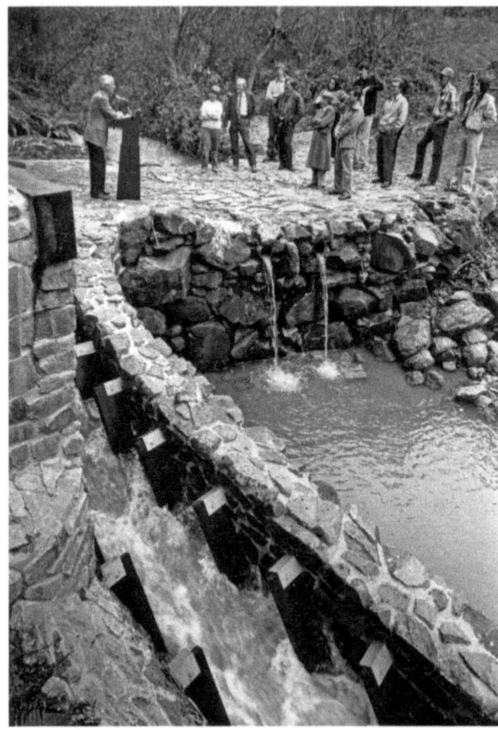

Park district director Harlan Kessel officiates at the dedication of the Denil fishway on May 6, 1987. The genetically unique rainbow trout swim between Lake Chabot and their higher spawning grounds. *Courtesy East Bay Regional Park District archives.*

The solution was to build a Denil fishway on the creek, about two hundred feet inside the park main entrance. Unlike a fish ladder, the Denil fishway is a thirty-foot ramp, constructed of native rock and redwood, through which the fish can swim without having to jump.

In the early spring, the fish emerge from upper San Leandro Reservoir, swim upstream, transit the fishway and return to gravel beds to spawn.

Public information officer Ned MacKay described the Redwood Park trout in a 1983 news release: "Adapting to new circumstances, they are believed to have survived and propagated, isolated from cross-breeding with other species and hatchery fish. This would make them one of the last native rainbow trout populations near a large urban area in California."[319]

Recognizing the historic and scientific importance of the Redwood Park rainbow trout population, district managers applied for and were granted historic landmark status for the fish population. A state historic plaque was dedicated at the fishway on May 6, 1987.[320]

By the 1990s, another threat had emerged to endanger the protected trout: off-leash dogs, in large numbers, were digging up the spawning areas in gravel beds. Park staff posted signs prohibiting people and dogs from going into the creek. Still, many visitors would not comply. By 2000, the park district Ordinance 38 was amended to require dogs to be on leash in the Stream Trail area. Public education to enforce this rule has been an ongoing effort for park staff and trail safety volunteers.

EPILOGUE

Our mission is to welcome and assist park users, and improve the quality and safety of Oakland's wildland park trails.
—Stan Dodson, Oakland Park Patrol

The coming of the twenty-first century has seen a great increase in the number of visitors coming to he East Bay redwood region.

With the invention of the mountain bicycle in the 1970s, hikers and equestrians began to object to difficult encounters with speeding bicyclists on trails. Hikers, particularly the elderly and people with disabilities, became frightened and were sometimes injured by bicycles. Horses spooked by bicycles sometimes threw and injured riders.

In a gesture of goodwill, organizations of mountain bicyclists participated in trail-maintenance activities and public outreach to improve their relationships with other user groups.

As the number of bicyclists grew, prevailing sentiment among this community called for access to all parkland trails, be they fire roads or single-track trails. Regional park policy generally defined the former as "multi-use" trails, with bicycles prohibited on the latter. In 2016, EBMUD directors began to investigate whether to allow bicycles on the watershed trails, in response to requests from the bicyclist community. Environmental, hiking and equestrian groups objected. The controversy is ongoing.

Dog walking has also become a very popular activity in the parklands. There have been cases where dogs pursued, and sometimes bit, bicyclists.

And there were cases where dogs attacked and injured horses, sometimes resulting in riders being thrown and injured.

An ongoing program of peer education was instituted in the East Bay hills, starting in about 2000. "Partners on the Trail," a collaborative program conducted by the Metropolitan Horsemen's Association and the Oakland Dog Owners Group, was held in Redwood Regional Park from 2001 through 2006, with support from park supervisor Dee Rosario.

Membership in the EBRPD Volunteer Trail Safety Patrols, with mounted, bicycle, hiking and companion dog units, increased. Trained volunteers wear uniforms and carry radios. They observe conditions in the parklands, report unlawful conduct and educate the public as to rules and safety conditions. "The volunteers are a 'force multiplier' for the park district police department," said EBRPD police chief Timothy Anderson.

Meanwhile, in Dimond, Leona and Joaquin Miller Parks in the city of Oakland, volunteer Stan Dodson in 2015 organized the Oakland Parks Patrol, modeled after its regional park district counterpart.[321] Dodson in 2014 produced a short film called *Trailhead*, in which community leaders describe what the parks mean to them personally, as well as how these organizations assist in bringing underserved youth and adults to the parks in the East Bay hills.

NOTES

Chapter 1

1. Saclan is pronounced "*sock* lawn." Huchiun is pronounced "hoo *chee* un." Jalquin-Yrgin is pronounced "haul keen *eer* gun."
2. The name *Costanoan*, "people of the coast," was given by the Spanish. It refers to the people who are now known as Ohlone.
3. Sandy Kimball and Brother Dennis Goodman, FSC, *Moraga's Pride: Rancho Laguna de los Palos Colorados* (Moraga, CA: Moraga Historical Society, 1987).
4. Donald D. Walker, "Moraga before 1900: A Community's History," unpublished thesis, California State University–Hayward, 1989.
5. Randall Milliken, *A Time of Little Choice: The Disintegration of Tribal Culture in the San Francisco Bay Area 1769–1810* (Banning, CA: Ballena Press, 2009), 157.
6. Ibid., 146.

Chapter 2

7. M.M. Wood, *History of Alameda County* (reprint, n.p.: Holmes Book Company, 1969; originally published in 1883).
8. Hubert Howe Bancroft, *Histories of Western North America*, vol. 34, *California Pastoral, 1769–1848* (San Francisco, CA: History Company, 1888), http://onlinebooks.library.upenn.edu/webbin/metabook?id=worksbancroft.
9. CSU Northridge California History Online, "Mexican California: The Heyday of the Ranchos," www.csun.edu/~sg4002/courses/417/readings/mexican.pdf.

10. Herbert E. Bolton, *Anza's California Expedition*, vol. 3 (page 136), vol. 4 (page 361) (Berkeley, CA, 1930). See also *Publications of the Academy of Pacific Coast History* 2, "Diary of Pedro Fages Expedition to San Francisco Bay in 1770" (1911): 12–13.
11. Historian Dennis Evanosky has attempted to locate the vantage point from which Font drew his map. It appears to have been in the Leona Heights area.
12. William Heath Davis, *Sixty Years in California: A History of Events and Life in California* (San Francisco, CA: A.J. Leary Publishers, 1889). Scanned image available online as a free e-book from the California Digital Library.
13. Ibid., re: rancheros' diet.
14. Cecilia Peña noted that women were highly educated in some of the *Californio* families, including her own.
15. Davis, *Sixty Years in California*, re: Indians.
16. Ibid., re: bear scare.
17. Wood, *History of Alameda County*, re: horsemanship.
18. Davis, *Sixty Years in California*, re: bull and bear fights.
19. Cecelia Peña, interview with the author, 2016.
20. Gloria E. Miranda, "Racial and Cultural Dimensions of Gente de Razón Status in Spanish and Mexican California," *Historical Society of Southern California Quarterly* 70 (1988).
21. Joseph Lamson, *Round Cape Horn: Voyage of the Passenger-Ship James W. Paige, from Maine to California in the Year 1852* (Bangor, ME: Press of O.F. and W.H. Knowles, 1878). Available as a free e-book from Project Gutenberg.
22. Erwin Gustuv Gudde, *California Place Names: The Origin and Etymology of Current Geographical Names* (Berkeley: University of California Press, 1960).
23. Davis, *Sixty Years in California*, re: Vincente Peralta.
24. William Heath Davis in 1838 recounted some of his conversations with the old don. Davis, *Sixty Years in California*.
25. Eugène Duflot de Mofras, *Travels on the Pacific Coast* (reprint, n.p.: Santa Ana Books, 1937).
26. In Spanish, "Nueva Helvetia." The colony was named for Sutter's native Switzerland.
27. Wood, *History of Alameda County*.
28. Squatters defense league manifesto, quoted by Wood, *History of Alameda County*.
29. Wood, *History of Alameda County*, re: league convention.
30. Lamson, *Round Cape Horn*.
31. Domingo's letter to Horace Carpentier, from letters of the Peralta family, translated by the Antonio Peralta Hacienda archives.

Chapter 3

32. Herbert Eugene Bolton, *Fray Juan Crespi: Missionary Explorer of the Pacific Coast 1769–1774* (Berkeley: University of California Press, 1927).
33. Dennis Evanosky, interview with the author, 2016.
34. Also known as Louis Choris.
35. Cormac F. Lowth, *Maritime Art and Duin Laoghaire*, an illustrated talk given to the Duin Laoghaire Historical Society on February 21, 2007, and to the Maritime Institute of Ireland on March 20, 2008.
36. The first season was November 6–December 28, 1826. The second was November 18–December 5, 1827.
37. *Social Studies Fact Cards: California Explorers: Frederick Beechey* (N.p.: Toucan Valley Publications, 2016).
38. Different sources quote various dimensions. These numbers were stated in the 1866 report by Heuer and Cordell of the Army Corps of Engineers.
39. Beechey; *Narrative of a Voyage to the Pacific.*
40. Alan Fraser Houston, "Cadwalader Ringgold, U.S. Navy," *California History* 79, no. 4 (Winter 2000): 208.
41. Ibid.
42. Ibid.
43. Dynamite was invented by Alfred Nobel the same year, 1867, but it had not come into widespread use.
44. Nilda Rego, "A Rock Called Blossom Was Hazard to Ships," *Contra Costa Times*, April 15, 2001, quoting material from 1943 Society of California Pioneers journal and 1866 annual report of the United States Coast Survey.
45. By some accounts, the city conducted a contest, offering $75,000 as a prize. This idea predated the modern practice of soliciting competitive bids but was essentially the same process.
46. Biography of Allexey W. Von Schmidt, in Glenn B. Ashcroft, *Society of Engineers Year Book* (N.p., 1928).
47. The name of the Latvian-born Von Schmidt is variously spelled Alexis, Allexey and Alexy. His daughter, Lilly, later married Charles Lee Tilden.
48. Rego, "Rock Called Blossom Was Hazard to Ships."
49. James R. Smith, *San Francisco's Lost Landmarks* (Fresno, CA: Craven Street Books, Linden Publishing, 2005).
50. Nilda Rego, "The Day They (Finally) Blew Up Blossom Rock," *Contra Costa Times*, April 22, 2001, quoting material derived from 1943 Society of California Pioneers journal and 1866 annual report of the United States Coast Survey.

51. Howard A. Kelly, J.H. Barnhart biography, in *A Cyclopedia of American Medical Biography*, vol. 1 (N.p.: W.B. Saunders Company, Publishers, 1920).
52. According to Office of Historic Preservation, California State Parks. The original name Gibbons gave the trout was *Salmo iridea*. It is now known as *Oncorhynchus mykiss irideus*.
53. William P. Gibbons, MD, "The Redwood in the Oakland Hills," *Erythea* (1893). This is a botanical journal of limited circulation. The quotation would seem to describe today's Big Trees area in Joaquin Miller Park.
54. The redwood stump that Gibbons measured as 33.5 feet wide, including the bark, would have towered 300 feet tall. He also noted a "prodigious triple trunk" that aggregated 57 feet in diameter.
55. Alfred Russel Wallace, *My Life: A Record of Events and Opinions* (New York: Dodd, Mead and Company, 1905).
56. Letter from William P. Gibbons to John Muir dated April 7, 1896, one of many letters exchanged by the two; it is included in the John Muir collected papers, Stuart Library of Western Americana, University of the Pacific, Stockton, California.
57. Gibbons, "Redwood in the Oakland Hills."
58. Ibid.
59. (Frederick J.) Monte Monteagle, "How Early-Day Alamedan Pioneered Conservation," *Alameda Times-Star*, December 16, 1970.
60. Ibid.
61. Sherwood D. Burgess, "Forgotten Redwoods of the East Bay," *California Historical Quarterly* 30, no. 1 (March 1951): 1–14.
62. The Strehlow family was prominent in the city of Alameda. From 1923 to 1939, they owned the Neptune Beach Amusement Park, the "Coney Island of the West," at the site of today's Crown Beach and Crab Cove.
63. Frederick William Beechey, *Narrative of a Voyage to the Pacific and Beering's Strait* (Philadelphia, PA: Carey and Lea, 1832).
64. John Nicoles, conversation with the author, 2016.
65. The John B. Dewitt Redwoods State Natural Reserve near the town of Redway in Humboldt County is named in his honor.
66. Nicoles, conversation with the author in 2016.

Chapter 4

67. Kimball and Goodman, *Moraga's Pride.*
68. Burgess, "Forgotten Redwoods of the East Bay."

69. K. Jan Oosthoek, professor of history, "The Role of Wood in World History," Australian National University, Canberra, October 15, 1998, www.eh-resources.org. The Spanish Armada was destroyed by the British navy, commanded by Sir Francis Drake, in 1588. This began the decline of Spanish naval supremacy. The English still had abundant forest resources with which to build ships.
70. Burgess, "Forgotten Redwoods of the East Bay."
71. The second right not granted was possession of the seaport.
72. *Exploration du Territoire de l'Orégon, des Californies et de la Mer Vermeille, Exécutée Pendant les Années 1840, 1841 et 1842*. See more at http://www.klinebooks.com/pages/books/35251/m-duflot-de-mofras-eugene/exploration-du-territoire-de-loregon-des-californies-et-de-la-mer-vermeille-executee-pendant-les#sthash.XUnuJEfT.dpuf.
73. Neil Havlik and John Nicoles, *Redwood Regional Park Resource Analysis*, accepted by EBRPD Board of Directors September 16, 1975, revised October 1977.
74. Some publications give his name as "James Lamson."
75. Frederick C. Monteagle, *A Yankee Trader in the California Redwoods* (Oakland, CA: East Bay Regional Park District, 1976), 4.
76. Lamson, *Round Cape Horn*, quoted in "Days Gone By" (column), "Joseph Lamson's Nine Years in Northern California, a Snapshot of Mid-19th-Century Life," *Contra Costa Times*, December 1, 2013.
77. Monteagle, *Yankee Trader in the California Redwoods*, part 1, 1.
78. *San Francisco Daily Evening News*, "Lynch Law in San Antonio!! Two Men Hung!!!," Wednesday, August 23, 1854.
79. *Seven Years Street Preaching in San Francisco* (reprint, Ann Arbor: University of Michigan Library, 1856), quoted by Gary Kamiya, "Portals of the Past" (column), *San Francisco Chronicle*, October 29, 1916.
80. See Burgess, "Forgotten Redwoods of the East Bay."
81. William Taylor, *California Life Illustrated* (N.p., 1860, reprinted by Google Books).
82. Monteagle, *Yankee Trader in the California Redwoods*.
83. Ibid.
84. Lamson, *Round Cape Horn.*
85. The traditional Hawaiian dress with a long train is called *holoku.*
86. Isabella Bird Bishop, *Six Months Among the Palm Groves, Coral Reefs, and Volcanoes of the Sandwich Islands* (N.p., 1875).
87. Lamson, *Round Cape Horn.*
88. Ibid.

89. See Barbara Lekish, *Embracing Scenes about Lakes Tahoe and Donner: Painters, Illustrators, & Sketch Artists (1855–1915)* (Lafayette, CA: Great West Books, 2003).

Chapter 5

90. "The Knave" (column), "Redwoods Atop Oakland Hills First Brought Settlers Here," *Oakland Tribune*; October 9, 1966.
91. Ibid.
92. The latter name is rooted in the lore of the Girl Scouts, who would come much later.
93. "The Knave," "Redwoods Atop Oakland Hills." Also according to a notation by Monteagle on a photo from the Knowland Collection in the Oakland Public Library.
94. Local historian Dennis Evanosky, Sierra Club hike leader Ron Ucovich and retired ranger Rich Wirkkala, having investigated likely sites along Palo Seco Creek, agree that this is the most likely location for the Palo Seco Mill site.
95. Monteagle, *Yankee Trader in the California Redwoods*, 6.
96. Burgess, "Forgotten Redwoods of the East Bay."
97. Monteagle, *Yankee Trader in the California Redwoods*, 8.
98. "Tres sendas" is Spanish for "three paths." Some early documents misspell it as "Tres Cendes."
99. Theodore Henry Hittell, *The General Laws of the State of California, from 1850 to 1864 Inclusive*, vol. 1 (N.p.: H.H. Bancroft and Company, 1870).
100. Burgess, "Forgotten Redwoods of the East Bay."
101. Wood, *History of Alameda County*.
102. "The Knave," "Redwoods Atop Oakland Hills."
103. *Redwood Regional Park Resource Analysis*, 1977.
104. Ibid.
105. Burgess, "Forgotten Redwoods of the East Bay."
106. The court case was *Brown and Plummer v. Lee*, Third District Court, no. 738, April 6, 1858.
107. Burgess, "Forgotten Redwoods of the East Bay."
108. "The Knave" (column), "Early Sawmills," *Oakland Tribune*, September 7, 1941.
109. Ibid.
110. William Taylor, the mill owner, was a different person from the Methodist missionary of the same name who preached the first sermon to the Redwood Boys.

111. Monteagle, *Yankee Trader in the California Redwoods*, 10.

112. Contra Costa Supervisors meeting minutes, vol. 1, page 112. Available through the Contra Costa Historical Society, Martinez. See also the county map of the *Ranchos of Vicente and Domingo Peralta*, filed on January 21, 1857.

113. Alameda County Supervisors minutes, vol. 1, page 169. Available through the Oakland History Room, Oakland Main Library.

114. EPRPD *Redwood Regional Park Resource Plan*; 1977.

115. Ibid.

116. "The Knave" (column), "Old Logging Road," *Oakland Tribune*, September 7, 1941.

117. Alameda County Supervisors minutes, vol. 1, pages 61 and 169.

118. John Wesley Noble, *Its Name Was MUD: The Story of Water* (Oakland, CA: East Bay Municipal Utility District, 1970).

119. EPRPD *Redwood Regional Park Resource Plan*; 1977.

120. The canyon and Thornhill Road were named after Hiram Thorn.

121. Indian Gulch is today's Trestle Glen district. This is not to be confused with Indian Valley, which is located northeast of the town of Canyon in Contra Costa County.

122. Wood, *History of Alameda County*.

123. John Nicoles, interview with the author, October 2016.

124. Burgess, "Forgotten Redwoods of the East Bay."

125. Monteagle, *Yankee Trader in the California Redwoods*. The absence of first-growth stumps was noted by Monteagle.

126. Gibbons, "Redwood in the Oakland Hills."

127. Leona Heights Park was named McCrea Park at the time Paul Covel discovered "Old Survivor."

128. *Oakland Tribune*, "Original Redwood Here: Gnarled Redwood Tree May Be Oldest Living Thing in Eastbay," Wednesday, August 13, 1969, 28-A.

Chapter 6

129. Wood, *History of Alameda County*.

130. Beth Bagwell, *Oakland: The Story of a City* (Oakland, CA: Oakland Heritage Alliance, 1982, 2012), 48. According to Bagwell, Carpentier claimed 519 votes, despite the fact that the 1852 census had counted only 150 residents.

131. Davis, *Sixty Years in California*.

132. According to Bagwell (page 36), the name was suggested by Thomas Eager, the sawmill owner, who had been elected supervisor of the township, in honor of the USS *Brooklyn*. This was the ship on which Eager had come to California, in one of the earliest shiploads of Yankee emigrants.
133. The exact location where the stream emanates from underlying rock is marked with a signpost in today's Joaquin Miller Park, at the picnic area below Skyline Boulevard and above the Sequoia Horse Arena.
134. "Sausal" is translated as "willow grove."
135. Lisa Owens-Viani, "The Sausal Creek Watershed: A Cultural and Natural History," unpublished study document done by the Aquatic Outreach Institute and Friends of Sausal Creek, Richmond, CA, 1998.
136. Ibid.
137. The name was spelled "Lewelling" when the family lived in Iowa. Geneological essays by descendants suggest that Henderson changed the spelling at the time he migrated west, due perhaps to a feud with other relatives. Henderson may also have adopted the middle name William at that time. The spelling "Luelling" appears on his tomb at the Mountain View Cemetery in Oakland.
138. This is according to the records of the Peralta Hacienda Historic Park and Museum. Erica Mailman credits Henderson Luelling himself; Lewelling family archives maintain that it was Alfred who came up with the name.
139. Jean Leeper, "Henderson Lewelling Family," Rootsweb, 2009, http://www.rootsweb.ancestry.com/~ialqm/HendersonLuelling.htm.
140. *Sacramento Daily Union*, "The Emigrant Free Lovers," May 19, 1860.
141. Michael Colbruno and Dennis Evanosky, *Lives of the Dead: Mountain View Cemetery in Oakland* (Oakland, CA: self-published by the blog of the same name, 2008).
142. "The California Reflections of Caspar T. Hopkins," 1888, reprinted in *California Historical Society Quarterly* 27, no. 1 (1948): 65–73.
143. Oakland Wiki, "Hugh Dimond," based on articles in the *San Francisco Call*, https://localwiki.org/oakland/Hugh_Dimond.
144. *Montclarion*, October 3, 1997, quoted by Eleanor Dunn.
145. This description of Duncan appears in the obituary notice for Alexander E. McNee in the *Oakland Tribune* on April 8, 1938. The same words are used in the *San Mateo Coast Sector Volunteer Study Guide* dated January 2009, http://static1.1.sqspcdn.com/static/f/1461275/21375302/1356726976897/SMCoast-StateParks.pdf?token=7zpQF01mdO3vOsnho0C17ELKcJQ%3D.

146. Peter Banks, "An Investigation of the Cultural Resources within the Anthony Chabot Regional Park, Alameda County, California," California Archeological Consultants Inc., prepared in September 1982 for the EBRPD. The park district boundary line between Redwood and Anthony Chabot Regional Parks is Redwood Road. Thus, this site is included in Chabot Park even though it is a short distance from the Redwood Park main entrance.
147. Joseph Lamson, from an 1853 diary entry, quoted by Monteagle, *Yankee Trader in the California Redwoods*, 53.
148. Dodie Livingston, "Door Closes Last Time at Tiny Redwood Canyon School," *Oakland Tribune*, June 12, 1964.
149. Bill Stroebel, "3 R's Fit Nicely in One Room," *Oakland Tribune*, October 2, 1956.
150. The apple and cherry trees by the schoolhouse were planted by John (Marcillio) Reis.
151. Livingston, "Door Closes Last Time."
152. According to John Whatley, the original house burned down, and his grandfather built another house in the same place. Later, after the park district acquired Redwood Canyon for parkland in 1939, John Reis moved to a house near Piedmont Stables, at 6301 Redwood Road.
153. Quotations from Lydia Reis Whatley are from her handwritten notes, provided courtesy of John Whatley.
154. Today's MacArthur Boulevard.
155. A precise inventory of John Reis's stonework is not possible. Information about which stone fireplaces and stairs he constructed, or may have constructed, has been derived from the legend and lore shared by longtime local residents, including Lon Gruenfeldt.
156. San Francisco florists who have been in business since 1871.
157. Lydia Reis Whatley, who lived nearby on Rishell Street, often came to visit her childhood home during the years when Dee Rosario was supervisor. She told him many details about early times in Redwood Canyon.
158. Also known as the Big Bear Inn.
159. *Oakland Tribune*, "Two Counties Launch New Gaming Raids," October 22, 1935.
160. Reported in the *Oakland Tribune* on July 4, 1936.
161. *Oakland Tribune*, "One Shot in Tavern Fight," January 27, 1938.
162. *New York Times*, "Turk Murphy to Perform at Carnegie," January 9, 1987.

163. Yerba Buena Jazz Band site, http://jazzhotbigstep.com/372012.html. Text by Dave Radlauer quotes clarinetist Bob Helm in this description.
164. Ibid., http://jazzhotbigstep.com/YBJB/Big_Bear_Stomp.mp3.
165. *Oakland Tribune*, "Co-Ed Loses $600 in Jewelry, Suitcase," April 14, 1947.
166. Across from Redwood Road signpost no. 241.
167. Erika Mailman, "Looking Back" (column), *Montclarion*, March 2007.
168. Dionisio "Dee" Rosario, in the course of a long career with the East Bay Regional Park District, was Redwood Park supervisor from 1996 until his retirement in 2013. In 2016, he was elected to the district board of directors for Ward 2, representing the region surrounding the East Bay redwood forest.
169. Robert Bruce Anderson, "Urban Conservation and Urban Design, San Francisco," *Historic Resource Evaluation of 6415 Redwood Road*, March 1994.
170. Published investigations of Canyon include John Van Der Zee, *Canyon: The Story of the Last Rustic Community in Metropolitan America* (New York: Harcourt, Brace, Jovanovich Inc., 1971); Annabelle Williams, *Handmade Lives* (Canyon, CA: Firefall Editions, 2002); Art Boericke and Barry Shapiro, *Handmade Houses: A Guide to the Woodbutcher's Art* (N.p.: A&W Publications, February 1973); Janet McEwan, "Canyon, a History 1840–1939: The Development of a Suburban Village in Contra Costa County," MA thesis, Holy Names College, July 1, 1973.
171. McEwan, "Canyon, a History."
172. Ibid., quoting Albert McCosker.
173. Williams, *Handmade Lives.*
174. According to an 1886 deed, cited by McEwan from Henry Bird to Henry S. Jones, *Deeds*, vol. 49, April 7, 1886, 327.
175. Gladys Shally, "Canyon Echoes," *Orinda-Moraga Record*, April 20, 1961.
176. McEwan, "Canyon, a History."
177. Ibid., quoting an interview with Eleanor Dickenson, postmaster of Moraga, July 21, 1972.
178. Gladys Shally, "When Redwoods Rang...," *Lafayette Sun*, August 31, 1951, 9.
179. Bird to Chase, Bill of Sale, *Miscellaneous*, vol. 3, March 11, 1889, 204.
180. McEwan, "Canyon, a History," 35.
181. Ibid., 36.
182. Shally, "Canyon Echoes."
183. McEwan, "Canyon, a History." She quotes longtime Canyon resident Beryl Burpee.

184. Note that the Redwood Inn in Canyon, on Pinehurst Road, is entirely different from the Redwood Inn located near the Montiero Ranch on Redwood Road to the south.
185. McEwan, "Canyon, a History," quoting an interview with Grace Siefker in Lafayette, California, on July 21, 1972.
186. McEwan interview with Beryl Burpee, one of those affected, in San Leandro, July 28, 1972.
187. Noble, *Its Name Was MUD.*
188. Van Der Zee, *Canyon.*
189. Ibid.
190. Williams, *Handmade Lives.*
191. Van Der Zee, *Canyon*, 37.
192. Ibid., 45.
193. Williams, *Handmade Lives.*
194. J.D. O'Connor, "The Fuel Line Bombing and Canyon Fire of 1969," *Lamorinda Patch*, February 19, 2012; see also Terry Ryan, Associated Press, March 17, 1969.
195. *Piedmont Pines Neighborhood Association Newsletter*, "History from Piedmont Pines Resident John Bouey," July 2015.
196. Ibid.
197. Williams, *Handmade Lives.*
198. For the 2012–13 school year, the most recent ranking data available, the Canyon K-8 school was given an "A" rating by the State of California.
199. Lloyd Graham Donaldson, interview with the author, March 8, 2016.
200. The East Bay Hills Project website on the history of the Sacramento Northern Railway is available at http://eastbayhillsproject.org.
201. The grade was 4 percent in some places.
202. Named for C.R. Wilcox, an early real estate speculator.
203. Harre W. Demoro, *California's Electric Railways* (Glendale, CA: Interurban Press, 1986).
204. Ibid.
205. McEwan, "Canyon, a History," quoting from an anonymous manuscript, "The Oakland, Antioch and Eastern," at the St. Mary's College archives.
206. McEwan, "Canyon, a History," quoting an interview with Vernon Sappers by Leonard Verbarg in "The Knave" column, *Oakland Tribune*, June 16, 1968: "Conrad Williamson, proprietor of the Redwood inn, wrote to the C.A. Hooper Company in 1916 requesting permission to put a merry-go-round…across the road from the Inn." Albert McCosker recalled having conducted the merry-go-round in a speech

presented to the Moraga Historical Society, at Del Rey School in Orinda, California, on November 23, 1967. The recording is on file in the St. Mary's College archive.

207. McEwan, "Canyon, a History." She learned about the parks through interviews in 1972 with longtime Canyon residents Gladys Shally and Al Weber. She also interviewed railroad expert Vernon Sappers on this topic.

208. McEwan, "Canyon, a History," citing the deed for "Oakland-Antioch, Oakland, Antioch and Eastern and San Ramon Valley Railroads to the San Francisco–Sacramento, *Deeds*, vol. 361, January 26, 1920, 2.

209. Ibid., quoting journalists and train buffs Harre DeMoro and Vernon Sappers on these points.

210. Ibid., quoting the correspondence of A.S. MacDonald and William G. Henshaw to Havens et al., *Assignment, Miscellaneous*, vol. 3, June 30, 1902, 516–17.

211. Garth G. Groff, "Havens, SN's Vital Middle of Nowhere," Sacramento Northern On-Line website, published by the Feather River Railroad Association, August 11, 2011, http://www.wplives.org/sn/havens.html.

212. One lot was not transferred to Ellis Wood—it was owned by Isaac (aka George) Williamson and was the site of the Redwood Inn.

213. McEwan, "Canyon, a History."

214. Van Der Zee, *Canyon.*

215. Dwayne McCosker, conversation with the author, March 2017.

216. Noble, *Its Name Was MUD.*

217. The stream that runs through Canyon is a tributary of San Leandro Creek.

218. Kimball and Goodman, *Moraga's Pride.*

219. The way the town name was pronounced by the old-timers was "val *vis*-ta," according to Dwayne McCosker, quoting his uncle, Joe Pereira.

220. Livingston, "Door Closes Last Time."

221. Banks, "An Investigation of the Cultural Resources."

222. Doris Marciel, conversation with the author, 2017.

223. The family name was originally Maciel. After 1920, the name was changed in a school record, adding an *r*, to become Marciel.

224. The gun range was closed by the park district in 2016 after fifty-two years in operation. The closure was in response to community concerns about lead pollution of groundwater, as well as complaints by visitors and area residents about the noise of gunfire.

225. To the left, as seen from the perspective of someone facing the targets.

226. The school was demolished and houses built on the site.

Chapter 7

227. Pauline Gershanov, "Biography of Joaquin and Juanita Miller," unpublished, March 17, 1979, Oakland History Room, Main Library.
228. M.M. Marberry, *Splendid Poseur: Joaquin Miller, American Poet* (New York: Thomas Y. Crowell Company, 1953).
229. Ibid.
230. Ed Henry, Shelley Rideout and Katie Wadell, *Berkeley Bohemia: Artists and Visionaries of the Early 20th Century* (N.p.: Gibbs Smith Publishers, 2008).
231. Dates are from a self-guided tour of the monuments written by Louis (Digger) O'Dell.
232. Their wedding was held "high in the Berkeley hills," according to the archives of the Contra Costa Hills Club.
233. Despite persistent rumors that Harold French himself cleared his eponymous trail by hand, no proof of this has come to light. French had a busy life. It is likely that the admirers of French did the trail work and had the trail named in his honor. Sidney Chown, however, did clear the trail in Redwood Regional Park that was later named after him.
234. Michael P. Cohen, *The History of the Sierra Club: 1892–1970* (San Francisco, CA: Sierra Club Books, 1988).
235. George Emanuels, "Duncan McDuffie: A Pioneer Conservationist," *House Key, News of Mason-McDuffie*, February 1977, courtesy of the corporate archives of the Better Homes and Gardens/Mason McDuffie company, Pleasanton, California. Emanuels (1908–2005) was a nephew of McDuffie's longtime business partner, Joseph Mason.
236. Steven Finacom, "Lecture Series by Phoebe Cutler Celebrates Artistry of Mills College Landscape," *Berkeley Daily Planet*, March 23, 2007.
237. *Oakland Tribune*, "Clubwomen Are Impressed by Trip to Heights and Havens Grove," 1922.
238. Louis Allen, "Saving Oakland's Sequoias: Eastbay Residents Have Grounds at their Disposal for Camps and Picnics," *Oakland Tribune*, February 24, 1924.
239. The area in question would appear to correspond to today's Roberts Regional Park. The chain of ownership of the area is not clear.
240. This is the site of today's Snow Park.
241. Allen, "Saving Oakland's Sequoias."
242. Oral history by Carol Sibley, quoted in Mimi Stein, *A Vision Achieved: Fifty Years of East Bay Regional Park District* (East Bay, CA: published by the district, 1984).

243. Frederick Isaac, *Jews of Oakland and Berkeley* (Charleston, SC: Arcadia Publishing, 2009). During the early twentieth century, the Kahn family was part of the downtown Oakland Jewish community and members of the Temple Sinai congregation who were committed to the betterment of the city. Founder Israel Kahn (1822–1883) was a pioneer Oakland merchant. The 1912 Kahn's Department Store building, at 1501 Broadway, designed by architect Charles W. Dickey, is landmark no. 89000194 in the National Register of Historic Places. It was restored during the 1990s, renamed the Rotunda Building and is now used as a business center. During the 2011 Occupy Oakland demonstrations, owner Phil Tagami notably guarded its doors against unruly Black Bloc anarchists, earning him the sobriquet "Shotgun Phil."

244. *San Francisco Examiner*, "Park Project Explained by Irving Kahn: Lake-Chabot-to-Richmond Beauty Spot Was Original Idea of Pioneer Benefactor," January 18, 1931.

245. Jack Turner, *Landscapes on Glass: Lantern Slides for the Rainbow Bridge—Monument Valley Expedition* (N.p.: Durango Herald Small Press, 2010). The author is a grandson of Ansel Hall.

246. Ibid.

247. Contra Costa County pulled out of the initiative, causing El Cerrito and Richmond to withdraw as well. Thirty years later, the political will would manifest itself for Contra Costa County to join the park district.

248. In 2008, during the Great Recession, bond issue AA for the district similarly won 71.7 percent of the votes cast.

249. Unwillingness of Contra Costa voters to tax themselves had caused the cities of El Cerrito and Richmond to drop out.

250. See Stein, *Vision Achieved*.

251. Elbert Vail, unpublished memoirs, available at the EBRPD archives.

252. An estimated three thousand workers were employed in the parklands by the WPA and CCC.

253. Letter dated November 15, 1934, to Edward M. Ament, Mayor of Berkeley, from Thomas J. Roberts, EBRPD archives.

254. *Oakland Tribune*, "Oakland Park Plan without Cost Proposed," September 15, 1935.

255. Laura McCreery, *Living Landscape: The Extraordinary Rise of the East Bay Regional Park District and How It Preserved 100,000 Acres* (Berkeley, CA: Wilderness Press, 2010).

256. Before the inception of the park district, voters had agreed to pay EBMUD a total of twenty-nine cents per $1,000 of assessed property

value to pay for the Pardee Dam construction on the Mokulumne River and to purchase the private water facilities in Contra Costa County. The 1936 land transfer agreement reduced the EBMUD tax rate from twenty-nine cents to twenty-five.

257. This agency was formed in 1933 with a mission to distribute state and federal funds to improve local conditions and relieve unemployment. It was superseded by the California State Relief Administration (SRA) in 1935.

258. Michael Hiltzik, *The New Deal: A Modern History* (New York: Simon and Schuster, 2011).

259. Ibid., 311.

260. Stein, *Vision Achieved.*

Chapter 8

261. Ibid.

262. Ibid.

263. Ibid.

264. Title search information is documented in a letter dated October 20, 1949, from Oakland Title Insurance and Guarranty Company to Andrew F. Burke, attorney for the Roman Catholic Archdiocese of San Francisco, concerning 58.34 acres being sold to the EBRPD.

265. *San Francisco Call*, "Dimond the Beautiful: Where Sylvan Charms and Agricultural Prosperity Meet," April 13, 1896.

266. Letter to Andrew F. Burke, attorney for the archdiocese, dated February 2, 1949, signed by Thomas J. Roberts, secretary of the EBRPD.

267. "The Knave" ran in the *Oakland Tribune* from 1933 until the death of publisher Joseph Knowland in 1974. Over its long run, the unsigned column was written by several reporters. It is most often associated with Leonard Verbarg, who wrote the column from 1954 to 1974. The column in which Dr. John Engs reminisces about his rides to the "Closson" farm was likely written by Addison "Ad" Brown Schuster, who is credited with authoring the column from 1938 to 1953.

268. No historical record has come to light indicating that there was any lumber mill located in or near the Redwood Bowl.

269. "The Knave" column, September 7, 1941.

270. The unattributed resource study is believed by Jerry Kent to have been written by John Nicoles and Neil Havlik.

271. Charles E. Snook (1863–1940) was a prominent Oakland lawyer and member of the Native Sons of the Golden West.
272. According to a 1934 map made by B.N. Flagg for Elbert Vail, showing the names of the owners of the land that would become Redwood Regional Park.
273. According to the 1878 Thompson and West *Alameda County Atlas.* The same atlas shows the lower portion of the Tate Trail, east of Piedmont Stables, as an access road to a residence near the Big Spring.
274. *Oakland Tribune*, "Shopping Tour Leads to Altar: Wealthy Bachelor to Wed Owner of Oakland Gift Store," February 10, 1938. See also *Oakland Tribune*, "Nephew Contests $100,000 Will of Alexander E. McNee," May 12, 1938.
275. From Roberts letter to Andrew Burke, attorney for the archdiocese.
276. According to the 1977 *Redwood Regional Park Resource Analysis*, page 6.
277. In a letter to the archdiocese from attorney Andrew F. Burke, October 20, 1949. Burke stated that Elsie Belle Brougher had a "stray interest" in the property. The attorney recommended that the parties undertake a "quiet title action."
278. The EBRPD defines a recreation area as fewer than one hundred acres.
279. Stein, *Vision Achieved.*
280. Hank Pellissier, "Children's Fairyland," *New York Times*, February 5, 2011. The writer notes that Disney toured Fairyland early after its opening on April 11, 1950, and created Disneyland in Anaheim soon thereafter. It hired away some of the Oakland Fairyland staff.
281. *Oakland Tribune*, "Children's Fairyland," April 12, 1950.
282. McCreery, *Living Landscapes*.
283. Lloyd Graham, in conversation with the author, 2016.
284. The City of Oakland maintained a mounted police unit until 2003.
285. Then located at 11500 Skyline Boulevard, Oakland.
286. Lloyd Graham Donaldson, interview with the author, March 8, 2016.
287. Carole Hicke, "Interview with Hulet Hornbeck," September 29, 1981, unpublished, EBRPD archives.

Chapter 9

288. Jessie Chown, Sid's mother, kept a diary of her life in Oakland, according to Heather Galanis of Palo Alto, granddaughter of Sidney Chown, in conversation with the author in 2013.

289. Arbor Villa stood south of today's Park Boulevard, on the hilltop between East Twenty-Second and East Twenty-Eighth Streets.
290. Franklin School was located at Ninth Avenue and East Eleventh Street.
291. Peggy Stinnett, "Horseman-Grocer Had a Wonderful Time in 75 Years," *Montclarion*, July 26, 1961.
292. The three-acre Chown ranch is now part of the Montclair district, but it was then in Piedmont. Thus, this was the original site of Piedmont Stables. The ranch is now a shopping center with Italian Colors restaurant and other businesses and medical offices.
293. Stinnett, "Horseman-Grocer Had a Wonderful Time."
294. Recollection of Heather Galanis.
295. Lucile Chown, *MHA Trail Blazer*, January 1962.
296. Florence also had a job for Planters Peanuts. She wore the "Mr. Peanut" costume at the 1939 World's Fair. During World War II, she worked for a radio station.
297. Stinnett, "Horseman-Grocer Had a Wonderful Time."
298. Chown, *MHA Trail Blazer*, January 1962.
299. The market remains at this location, with the street address 1440 Leimert. It has been known for many years as "Rocky's Market," having experienced many changes of ownership.
300. George S. Daniels, MHA past president, "Gone but Not Forgotten," *MHA Trail Blazer*, January 1952.
301. Ibid.
302. Ibid.
303. Amelia Sue Marshall and Terry Tobey, *Oakland's Equestrian Heritage* (Oakland, CA: Arcadia Publishing, 2007).
304. According to Bob Schultz, who worked with Scotty Black for the City of Oakland Parks Department, Scotty's ranch was located at 11500 Skyline Boulevard. The ranch was demolished to build a new park headquarters. The building is now called the Trudeau Training Center.
305. Now the East Hills Community Church property, previously called the Melrose Baptist Church.
306. Now the Pinto Playground and Carl Munck School.
307. Marshall and Tobey.
308. Bats continue to live in two park district horse barns—Piedmont Stables and Sunol—where they receive regular visits from wildlife biologists.
309. Yvonne White, conversation with author, 2007.

310. Some of this information is documented in the legal case *K.E. Bemis et al., Appellants, v. THE PEOPLE*, civil case no. 14772, First District Court, Division 1, February 15, 1952.
311. The well remains, a hole in the ground concealed by vegetation.
312. To accommodate the neighbors on Campus Drive, the park district has installed a large gate, locked from the outside, to prevent the public from entering Leona Canyon from the upscale neighborhood.
313. The late Jan Studley Hansen, who grew up at Leona Stables, conversations with the author, 2006 and 2007.
314. Dee Rosario and Annie Kenny, conversations with the author in 2016.
315. "Regional Park District to Acquire Key Property for Leona Heights," news release, November 9, 1998, East Bay Regional Park District.

Chapter 10

316. Gordon Laverty and Larry Laverty, "Leona Heights Neighborhood News," *MacArthur Metro*, April 2011.
317. Burress also illustrated Covel's second book, *Beacons Along a Naturalist's Path Trail: California Naturalists and Innovators* (N.p.: Western Interpretive Press, 1988).
318. *Oakland Tribune*, "Original Redwood Here," August 13, 1969, 28-A.
319. Ned MacKay, news release, March 9, 1983, EBRPD archives.
320. Ibid., May 6, 1987, EBRPD archives.

Epilogue

321. See Oaklandtrails.org.

INDEX

C

D

E

F

G

H

J

K

L

M

N

O

P

R

S

ABOUT THE AUTHOR

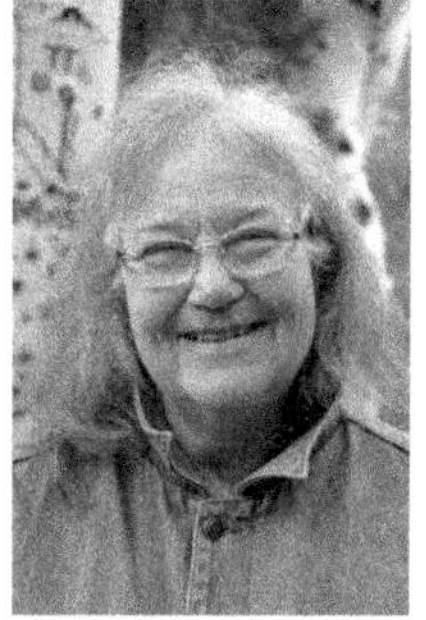

Amelia Sue Marshall is a writer who lives with her family on Peralta Creek in Oakland. She developed a passionate interest in the history of the East Bay redwood region while riding her mare there as a trail safety patrol volunteer for the East Bay Regional Park District. Her prior careers as a journalist, space science engineer, planetarium lecturer and real estate broker have been cast aside in favor of writing about local history. In addition to serving as a trail safety volunteer, she is a board member of the Metropolitan Horsemen's Association and the Oakland Heritage Alliance.

Visit us at
www.historypress.net

www.ingramcontent.com/pod-product-compliance
Lightning Source LLC
LaVergne TN
LVHW052339100826
845147LV00021B/1117

* 9 7 8 1 4 6 7 1 3 7 2 5 6 *